Basic Offshore Safety

T0144146

- Comprehensive insight into the offshore oil and gas industry for those intending to choose it as a career.
- Full syllabus coverage for OPITO's BOSIET, FOET, MIST and IMIST courses.
- Produced in full colour with over 180 images.

Basic Offshore Safety covers everything that newcomers to the offshore oil and gas industry need to know prior to travelling offshore or when attending OPITO's Basic Offshore Safety Induction and Emergency Training (BOSIET), Minimum Industry Safety Training (MIST), Further Offshore Emergency Training (FOET) and International MIST courses. Primarily focused on the oil industry, this book introduces readers to the key safety topics in the offshore support vessel industry and common to the renewable industry. Written in easy to follow steps and including references to both the legislation and guidance where relevant, Abdul Khalique walks the reader through the hazards they are likely to encounter when travelling to, from or working offshore, showing how to minimise risks and deal with any issues that may arise at any stage of the work.

Abdul Khalique is an experienced lecturer and manager in the field of maritime and offshore safety training. His experience includes several years of seafaring as a deck officer followed by a move to Warsash Maritime Academy as Principal Lecturer for the Maritime and Offshore Safety section, later becoming Senior Lecturer and Deck Cadet Programme Leader at the Shetland School of Nautical Studies. He currently works at Serco Marine Services as a Marine Training Officer.

Basic Offshore Safety

Safety induction and emergency training for new entrants to the offshore oil and gas industry

Abdul Khalique

Routledge
Taylor & Francis Group

LONDON AND NEW YORK

First published 2016
by Routledge
2 Park Square, Milton Park, Abingdon, Oxon OX14 4RN

and by Routledge
711 Third Avenue, New York, NY 10017

Routledge is an imprint of the Taylor & Francis Group, an informa business

© 2016 Abdul Khalique

British Library Cataloguing-in-Publication Data
A catalogue record for this book is available from the British Library

Library of Congress Cataloging in Publication Data
Khalique, Abdul.
Basic offshore safety : safety induction and emergency training for new entrants to the offshore oil and gas industry / Abdul Khalique.
pages cm
1. Offshore oil well drilling--Safety measures. 2. Offshore gas well drilling--Safety measures. I. Title.
TN871.3.K485 2016
622'.338190683--dc23
2015018917

ISBN: 978-1-138-84591-6 (pbk)
ISBN: 978-1-315-72782-0 (ebk)

Typeset in Sabon by
Servis Filmsetting Ltd, Stockport, Cheshire

Printed by Bell and Bain Ltd, Glasgow

MIX
Paper from
responsible sources
FSC® C007785

Contents

About the Author

Abdul Khalique Master Mariner, MICS, PG Cert. (Shipping), PG Cert. (TQFE), MSc (CBIS), HND Nautical Science, BSc (Maritime Studies).

Currently employed at Serco Marine Services as Marine Training Officer, Abdul Khalique began his sea career at Pakistan Marine Academy where he was awarded the President of Pakistan Gold Medal for the best cadet in BSc Maritime Studies. Later on he received the High Achievement Award and the Merchant Navy Association (Tasmania) prize on completion of the Advanced Certificate in Marine Operations (second mate) from the Australian Maritime College.

After obtaining an HND in Nautical Science in 2000 from Blackpool and the Fylde College, Fleetwood, United Kingdom, he sailed briefly and then continued his studies to earn an MSc in Computer Based Information Systems from the University of Sunderland. He then studied for the Postgraduate Certificate in Shipping at the New Zealand Maritime School. During his service at sea, he experienced working on several types of vessels including bulk, container, Ro-Pax and offshore support vessels.

In 2003, Abdul moved to the Shetland School of Nautical Studies at the NAFC Marine Centre as an HND Nautical Science course developer. He continued in this position until the Centre started training cadets in 2004 when he became a lecturer in Nautical Studies. While in this position, he passed the Institute of Chartered Ship Brokers examination. Continuing with his teaching profession, he also obtained a Postgraduate Certificate in Teaching Qualification for Further Education (TQFE) from the University of Dundee. Later in 2006, he was promoted to Senior Lecturer and Cadet Programme Leader for HND Nautical Science where he worked until September 2008 when he moved to Warsash to take up the Principal Lecturer, Maritime & Offshore Safety position at Warsash Maritime Academy (Southampton Solent University). Warsash provided him with an opportunity to expand his expertise in the offshore sector covering oil and gas as well as renewable industries. In June 2014, he moved to Serco Marine Services where he provides expert training advice to shipboard staff serving on a large variety of vessels.

Acronyms and Abbreviations

°C	Degrees Centigrade
2D	2 Dimensional
3D	3 Dimensional
AAD	Advanced Anomaly Detection
AAIB	Air Accident Investigation Branch
ACAS	Airborne Collision Avoidance System
ACoP	Approved Code of Practice
AFFF	Aqua Film Forming Foam
AFP	Active Fire Protection
AHTS	Anchor Handling Tug Supply Vessels
AIMS	Asset Integrity Management System
AIS	Automatic Identification System
AIT	Auto Ignition Temperature
ALARP	As Low as Reasonably Practicable
ANO	Air Navigation Order (CAA)
AOC	Air Operator's Certificate
APF	Assigned Protection Factor
AR-AFFF	Alcohol Resistant – Aqueous Film Forming Foam
ARCC	Aeronautical Rescue Coordination Centre
ASA	Advanced Safety Auditing
ASV	Accommodation Support Vessel
ATC	Air Traffic Control
ATPL	Airline Transport Pilot Licence
BA	Breathing Apparatus
BBS	Behaviour Based Systems
BCAS	Beacon Collision Avoidance System
BLEVE	Boiling Liquid Expanding Vapour Explosion
BLO	Boat Landing Officer
BOER	Basic Onshore Emergency Response
BOP	Blowout Preventer
BOSIET	Basic Offshore Safety Induction and Emergency Training
BP	British Petroleum
BS	British Standard
BSI	British Standards Institution
BSS	Behavioural Safety Systems or BBS Behaviour Based Systems
CAA	Civil Aviation Authority
CA-EBS	Compressed Air Emergency Breathing System
CALM	Catenary Anchor Leg Mooring
CAP	Civil Aviation Publication
CASOC	California Arabian Standard Oil Company
CBA	Cost Benefit Analysis

CCTV   Close Circuit Television
CE   Conformité Européenne – Mandatory Conformity marking for products sold in Europe
CFR   Code of Federal Regulations (USA)
CGBS   Concrete Gravity Based Structures
CHIP   Chemicals (Hazard Information and Packaging for Supply) Regulations, 2009
CLP   European Regulation (EC) No 1272/2008 on classification, labelling and packaging of substances and mixtures
cm   Centimetre
CO   Carbon Monoxide
CO_2 Carbon Dioxide
COSHH   Control of Substances Hazardous to Health
COSPAS   Space System for Search of Distress Vessels (Russian Acronym for)
CPA   Closest Point of Approach
CPL   Commercial Pilot Licence
CPR   Cardiopulmonary Resuscitation
CRO   Control Room Operator
CRT   Cathode Ray Tube (Also CRT – Coastguard Rescue Team)
CSS   Coordinator Surface Search
CTS   Carpal Tunnel Syndrome
dB   Decibels
DCR   Offshore Installations and Wells (Design and Construction) Regulations, 1996
DLR   Davit Launched Liferafts
DECC   Department of Energy and Climate Change (UK)
DES   Drilling Equipment Set
DfT   Department for Transport (UK)
DNV   Det Norske Veritas (Classification Society)
DP   Dynamic Positioning
DPO   Dynamic Positioning Officer
DORIS   Dropped Object Register of Incidents and Statistics
DSC   Digital Selective Calling
DSEAR   Dangerous Substances and Explosive Atmospheres Regulations, 2002
E&P   Exploration and Production
E/C&I   Electrical, Control and Instrumentation
EASA   European Aviation Safety Agency
EAV   Exposure Action Value
EBA   Escape Breathing Apparatus
EBS   Emergency Breathing System
ECC   Emergency Control Centre
ECT   Escape Chute Training
ECTR   Escape Chute Training Refresher
ED   Energy Division (HSE)
EEBD   Emergency Escape Breathing Device
EFIS   Electronic Flight Instrument Systems
EGC   Enhanced Group Calling
EHEST   European Helicopter Safety Team (Member of the IHST)
ELSA   Emergency Life Support Apparatus
ELT   Emergency Location Transmitter

ELV   Exposure Limit Value
EMS   Environmental Management System
EPA   Environment Protection Act, 1990
EPIRB   Emergency Position Indicating Radio Beacon
EPOL   Emergency Preparedness Offshore Liaison Group (UK)
ERCB   Energy Resources Conservation Board (Canada)
ERP   Emergency Response Plan
ERRV   Emergency Response and Recovery Vessel
ERS   Emergency Response Strategy
ERT   Emergency Response Team
ETSO   European Technical Standard Order
ETV   Emergency Towing Vessel
EU   European Union
EUBS   Emergency Underwater Breathing System
FAA   Federal Aviation Administration (USA)
FES   Fire and Explosion Strategy
FFA   Fire Fighting Appliance(s)
FFFP   Film Forming Fluoroprotein
FOER   Further Onshore Emergency Response
FOET   Further Offshore Emergency Training
FP   Fluoroprotein
FPD   Fall Prevention Device
FPS   Floating Production System
FPSO   Floating Production Storage and Offloading
FRC   Fast Rescue Craft
FSO   Floating Storage and Offloading
FTsbB   Further Training Travel Safely by Boat
GA   General Alarm (also General Arrangement when referred to ship's plans)
Galileo European Satellite Positioning System
GCSE   General Certificate of Secondary Education
GHz   Gigahertz
GLONASS Globalnaya navigatsionnaya sputnikovaya Sistema (Russian Positioning
         System)
GMDSS   Global Maritime Distress and Safety System
GMP   Garbage Management Plan
GNSS   Global Navigation Satellite System
GPA   General Platform Alarm
GPS   Global Positioning System
GRP   Glass fibre Reinforced Polyester (material)
gt   Gross Tonnage
H&S   Health and Safety
H_2O Water
H_2S Hydrogen Sulphide
HASAWA   Health and Safety at Work etc. Act 1974
HAVS   Hand-Arm Vibration Syndrome
HazID   Hazard Identification
HCA   Helideck Certification Agency
HCR   Hydrocarbon Release

HDA	Helideck Assistant
HEBE	Helicopter Emergency Breathing Equipment
HEED	Helicopter Emergency Egress Device
HERT	Helideck Emergency Response Team
HERTL	Helideck Emergency Response Team Leader
HERTM	Helideck Emergency Response Team Member
HLB	Hyperbaric Lifeboats
HLO	Helideck Landing Officer
HMCG	Her Majesty's Coastguard (UK)
HOMP	Helicopter Operations Monitoring Programme
hPa	Hecto Pascals (unit for measuring atmospheric pressure)
HRU	Hydrostatic Release Unit
HSE	Health and Safety Executive (UK)
HSW	Health and Safety at Work etc. Act (UK)
HUEBA	Helicopter Underwater Emergency Breathing Apparatus
HUET	Helicopter Underwater Escape Training
HUMS	Health and Usage Monitoring System
HVAC	Heating, Ventilation and Air-conditioning
Hz	Hertz
IAMSAR	International Aeronautical and Maritime Search and Rescue
ICAO	International Civil Aviation Organization
ICDS	Integrated Cockpit Display System
IFR	Instrument Flight Rules
IGS	Inert Gas System
IMIST	International MIST
IMO	International Maritime Organization
IOGP	International Association of Oil & Gas Producers
IR	Intervention Recommendations (also Implementing Rule for EASA)
ISO	International Standards Organization
JSA	Job Safety Analysis
kg	Kilogram
kN	Kilonewton
KPI	Key Performance Indicator
LAE	Licensed Aircraft Engineer
LAPP	Lifejacket Air Pocket Plus
LCD	Liquid Crystal Display
LEL	Lower Explosive Limit
LEV	Local Exhaust Ventilation
LOLER	Lifting Operations and Lifting Equipment Regulations (LOLER) 1998
LOP	Local Operating Procedure
LOTO	Lock Out Tag Out
LPG	Liquefied Petroleum Gas
LRRS	Lifeboat Release and Retrieval System
LSA	Life Saving Appliance(s)
LTA	Lost Time Accident
LTAR	Lost Time Accident Rate
MAC	Mid Air Collision
MAH	Major Accident Hazard

MARPOL	International Convention for the Prevention of Pollution from Ships, 1973/78
MCA	Maritime and Coastguard Agency (UK)
MEL	Minimum Equipment List
MEMIR	Major Emergency Management Initial Response
MES	Marine Evacuation System
MEWP	Mobile Elevating Work Platform
mg	Milligram
MGN	Marine Guidance Note
MHOR	Manual Handling Operations Regulations, 1992
MHz	Megahertz
MIST	Minimum Industry Safety Training
ml	Millilitre
MMEL	Master Minimum Equipment List
MOB	Man Overboard
MODU	Mobile Offshore Drilling Unit
MOU	Memorandum of Understanding
MPA	Maritime Patrol Aircraft
MRCC	Maritime Rescue Coordination Centre
MRSC	Maritime Rescue and Coordination Sub-centre
MSDS	Material Safety Data Sheets
MSF	Module Support Frame
MSLD	Maritime Survivor Locating Device
N_2O	Nitrous Oxide
NAF	Neal Adams Fire-fighters (USA)
NDT	Non-Destructive Testing
NOGEPA	Nederlandse Olie en Gas Exploratie en Productie Associatie
NORM	Naturally Occurring Radioactive Materials
Norsk Olje & Gas	Norwegian Oil and Gas Association
NOx	Nitrogen Oxide
NSOC-D	North Sea Operators Committee (Denmark)
NTO	National Training Organisation
NUI	Normally Unattended/Unmanned Installation
OEL	Occupational Exposure Limit
OEM	Original Equipment Manufacturer
OERTL	Offshore Emergency Response Team Leader
OERTM	Offshore Emergency Response Team Member
OERWG	Offshore Emergency Response Working Group
OGP	Now IOGP – International Association of Oil & Gas Producers
OIM	Offshore Installation Manager
OLF	Norsk Olje & Gas (Norway)
OLRRS	On-load LRRS
OODTP	Ongoing On-board Development & Training Programme for ERRV Masters and Crews
OOGUK	Oil & Gas UK
OPEC	Organization of Petroleum Exporting Countries
OPITB	Offshore Petroleum Industry Training Board
OPITO	Originally acronym for 'Offshore Petroleum Industry Training Organisation' but now used as a proper noun

OPPC   Offshore Petroleum Activities (Oil Pollution Prevention and Control) Regulations, 2005
OSC   On Scene Commander
OSD   Offshore Safety Division (HSE)
OSDR   Offshore Safety Directive Regulator
OSHA   Occupational Safety and Health Administration (US)
OSPAR   Convention for the Protection of the Marine Environment of the North-East Atlantic, 1992
OSPRAG Oil Spill Prevention and Response Advisory Group
OSV   Offshore Supply Vessel
PA   Public Address (System)
PAPA   Prepare to Abandon Alarm
PDCA   Plan, Do, Check, Act
PELB   Partially Enclosed Lifeboat
PFEER   Prevention of Fire and Explosion, and Emergency Response Regulations, 1995
PFP   Passive Fire Protection
PIC   Pilot in Command
PITB   Petroleum Industry Training Board
PLB   Personal Locator Beacon
POB   Persons On Board
POWER   Positive Observations Will Eliminate Risk (Chevron)
PPE   Personal Protective Equipment
PPL   Private Pilot Licence
ppm   Parts per million
PPM   Positive Performance Measure
PRfS   Personal Responsibility for Safety
P-STASS Passenger Short Term Air Supply System
PSV   Platform Supply Vessel
PTF   Petroleum Training Federation
PTW   Permit to Work
PUWER   Provision & Use of Work Equipment Regulations, 1998
RA   Resolution Advisories
RAF   Royal Air Force (UK)
RCC   Rescue Coordination Centre
REACH   Registration, Evaluation, Authorisation and Restriction of Chemicals
RIB   Rigid Inflatable Boat
RIDDOR Reporting of Injuries, Diseases and Dangerous Occurrences Regulations, 1995
RN   Royal Navy (UK)
RNLI   Royal National Lifeboat Institution
ROV   Remotely Operated Vehicle
RPE   Respiratory Protection Equipment
RSA   Radioactive Substances Act, 1993
SADIE   Safety Alert Database and Information Exchange
SALM   Single Anchor Leg Mooring
SAP   Systems Application Products
SAR   Search and Rescue
SARSAT Search and Rescue Satellite Aided Tracking.

SART   Search and Rescue Radar Transponder (Also Search and Rescue Satellite Aided Tracking)
SC   SAR Coordinator
SCBA   Self Contained Breathing Apparatus
SCE   Safety Critical Element
SCR   Offshore Installations (Safety Case) Regulations, 2005
SCUBA   Self Contained Underwater Breathing Apparatus
SDS   Safety Data Sheets
SI   Statutory Instrument
SIMOPS   Simultaneous Operations
SINTEF   Stiftelsen for Industriell og Teknisk Forskning (Offshore Blowout Database)
SMC   SAR Mission Coordinator
SMS   Safety Management System
SOC   Safety Observations and Conversations
SOLAS   International Convention on Safety of Life at Sea
SOP   Standard Operating Procedure
SOS   Safety Observation System (also Save Our Souls)
SOx   Sulphur Oxide
SPHL   Self-Propelled Hyperbaric Lifeboats
SPM   Single Point Mooring
SRG   Safety Regulation Group (CAA)
SRU   SAR Unit
SSD   Secondary Securing Device
SSOW   Safe System of Work
STASS   Short Term Air Supply System
STCW   International Convention on Standards of Training, Certification and Watchkeeping, 1978
STFs   Slips, Trips and Falls
STOP™   DuPont's Safety Training Observation Program
SUM   Structural Usage Monitoring
SWL   Safe Working Load
SWOT   Strengths, Weaknesses, Opportunities and Threats
TA   Traffic Advisories
TAD   Tender Assisted Drilling
TAWS   Terrain Awareness Warning System
T-BOSIET   Tropical Basic Offshore Safety Induction and Emergency Training
TCAS   Traffic Collision Avoidance System
TELB   Totally Enclosed Lifeboat
TEMPSC   Totally Enclosed Motor Propelled Survival Craft
T-FOET   Tropical Further Offshore Emergency Training
T-HUET   Tropical Helicopter Underwater Escape Training
TLP   Tension Leg Platform
TOFS   Time Out for Safety
TPA   Thermal Protective Aid
TRA   Task Risk Assessment
TsbB   Travel Safely by Boat
TUM   Transmission Usage Monitoring
UAE   United Arab Emirates

UEL	Upper Explosive Limit
UK	United Kingdom
UKCS	UK Continental Shelf
UKSRR	UK Search and Rescue Region
UN	United Nations
US	United States (of America)
USA	United States of America
USSR	Union of Soviet Socialist Republics
VD	Vapour Density
VFR	Visual Flight Rules
VHF	Very High Frequency
VVC	Vapour Vacuum Compression
WAH	Working at Height
WAHR	Work at Height Regulations, 2005
WEL	Workplace Exposure Limit
WOAD	World Offshore Accident Database
WP	Work Procedure

1

Offshore Oil and Gas

1.1 Introduction to the Petroleum Industry

The word 'petroleum' is made up of two Latin words, 'petra' meaning rock and 'oleum' meaning oil. In essence, petroleum is the liquid compound of hydrocarbons found beneath the surface of the Earth. This compound can be refined to produce petrol, diesel oil, paraffin and other petrochemicals.

Hydrocarbons, as the name suggests, consist of various proportions of hydrogen and carbon atoms. When found in liquid form, these hydrocarbons are referred to as 'crude oil' which varies in colour and viscosity due to its composition. The colours could be various shades of yellow or black. Oil consists of a mixture of liquid hydrocarbons[1] whereas gas consists mainly of methane but mixed with carbon dioxide, nitrogen, and helium or hydrogen sulphide. Since gas is lighter than oil, it is found prior to the discovery of oil. Due to high temperatures, some fossils convert into coal whereas the gases and liquids escaping this process form natural (and other gases) and oil. This process has been known to take 10–100 million years and continues even today.

Oil as a substance has been known to humankind for thousands of years. In the early 1840s, it was discovered in sand, springs and oil/tar pits and was used for lighting, street paving, medicine and waterproofing. As the world population and therefore the use of oil increased, processes were developed to refine kerosene from crude oil. The first land based well to facilitate the commercial supply

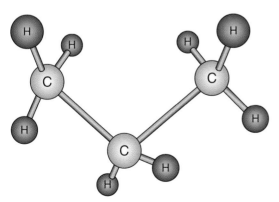

Figure 1.1 Propane (C_3H_8) molecule

Figure 1.2 Drilling and production piers at Summerland, California, USA (Source: http://aoghs.org/offshore-oil-and-gas-history/offshore-oil-history/)

of oil was drilled near Baku, Caspian Sea in 1848 and in Poland in 1854. In the 1850s and 1860s, Azerbaijan supplied up to 90 per cent of the oil supply to the world.

The presence of oil was known in North West Pennsylvania, USA for a long time. In 1859, the first well that produced ten barrels/day was drilled in Titusville, Pennsylvania. This find led to Pennsylvania becoming the lead oil supplier until 1901 when the largest production well was drilled in Spindletop Hill, south of Texas, USA. This well produced in excess of 100,000 barrels of oil per day.

The first offshore oil wells were drilled in around 1891 in Grand Lake St Marys, Ohio, USA. About five years later, oil wells in the Summerland field in California were drilled. The technique used at the time was very basic – wooden piers extending 400 metres from the shore with piles of 10 metres length reaching the seabed. Steel pipes were then pounded for depths of about 40 metres to reach the oil reserves giving a very modest yield and leaving enormous spillage into the sea.[2] By 1910, shore based drilling had begun in Iran, Sumatra, Venezuela, Peru and Mexico. The Taxas Company, which subsequently changed its name to Texaco and then to Chevron, developed the first mobile drilling units for the Gulf of Mexico.

Oil exploration commenced in 1933 in Saudi Arabia when oil concessions were granted to the Standard Oil Company of California which became the California Arabian Standard Oil Company (CASOC). Saudi Arabian oil reserves were subsequently discovered in 1938 to produce oil at commercial scale. In 1948, the largest oil field in the world, Ghawar, Saudi Arabia, was discovered.

In 1947, an offshore oil well was drilled in water depth of approximately 5 metres and 10.5 miles off the coast of Louisiana, USA. By this time, the technology had advanced to some extent whereby unidirectional pile drivers were used instead of rotary rigs. Instead of using wooden structures,[3] steel was used and barge drilling practice was put in place. The developments in offshore drilling were halted in 1950 by disputes of jurisdiction over the continental shelf and therefore leasing of the submerged lands around the coast of the USA. However, when the US federal government passed the 'Outer Continental Shelf Lands Act of 1953', they resumed leasing activity allowing further oil exploration.

In 1954, Mr Charlie, the first purpose built Mobile Offshore Drilling Unit (MODU), was used by Shell Oil to explore for oil in the Gulf of Mexico, USA. In 1961, the first four column semi-submersible drilling rig, owned and operated by the Blue Water Drilling Company, was deployed in the Gulf of Mexico for Shell Oil. In 1965, the British drilling barge 'Sea Gem' was the first rig to discover hydrocarbons 42 miles off the River Humber in the United Kingdom. Prior to it, a number of attempts had been made without much success. Up to a depth of 2500 metres, Sea Gem's drilling did not produce any positive prospects of finding oil or gas but after carrying out further tests, British Petroleum (BP) decided to continue drilling. Eventually, at a depth of 3000 metres a large reservoir of natural gas was discovered. This discovery resulted in a yield of 300,000 cubic metres of natural gas per day, a quantity that easily justified commercial operations. Unfortunately, when attempts were being made to relocate Sea Gem for drilling a second well, the rig suffered damage to two of its eight legs, resulting in the rig sinking in December 1965 causing a loss of 13 lives. The loss of Sea Gem resulted in a requirement to have a 'standby' vessel around the rig and an Offshore Installation Manager (OIM) nominated.

In 1960, Saudi Arabia, Kuwait, Iraq, Iran and Venezuela formed OPEC (Organization of Petroleum Exporting Countries). Later on, countries including Qatar, Indonesia, Libya, United Arab Emirates, Algeria, Nigeria, Ecuador, Angola and Gabon became members. In 1969, the offshore petroleum industry suffered its first major oil spill disaster when a platform 6 miles off the cost of Summerland USA suffered a natural gas blowout from a 1000 metres deep well. A 800 square mile slick of oil spilled into the Santa Barbara Channel, California, the equivalent of 80,000 barrels of oil leakage in a period spread over 11 days. This was the first major drilling incident in the history of oil and gas exploration. Consequently, the US government reviewed its environmental and regulatory policies for this type of operation.

Although oil production in the USA peaked in 1970, the country's dependency on oil created a shortfall leading to increased demand for oil from Arab nations. However, during the Arab–Israel conflict in 1973, OPEC imposed an embargo on oil exports to the USA, Netherlands, Portugal and South Africa, countries that supported Israel. This embargo was finally lifted in 1974 but in the process, the West changed its policy towards oil exploration. The Iranian revolution in 1979 caused the second oil crisis due to the reduction in oil production. Subsequently, in 1980 the oil production in both Iran and Iraq significantly reduced due to the Iran–Iraq war. During this time Venezuela, Mexico, Nigeria and the USSR increased their production and therefore their oil exports.

Due to increased depths and distance from shore, the exploration and production of offshore oil and gas present bigger challenges in comparison with land based drilling due to the remoteness and harshness of the environment. Whilst in the USA, Santa Barbara remained the biggest 'blowout' incident, similar incidents took place in the North Sea, Persian Gulf, Niger Delta and the Mexican waters of the Gulf of Mexico. The most recent major oil spill incident at the time of writing this book is the 'Deepwater Horizon Oil Spill' (April 2010) in which approximately 4.9 million barrels of oil were spilled – the largest oil spill in the history so far. The leaking well was finally sealed in September 2010 but by this time, the incident had caused the death of 11 employees with a US$37.2 billion estimated spill related cost.

Whilst on-board drilling rigs or production platforms, accidents can be the cause of a major loss of life or property and damage to the environment, many accidents take place during the transportation of personnel and equipment to and from offshore drilling rigs or production platforms. Table 1.2 gives a glimpse of helicopter accidents when travelling to/from offshore installations. Generally, this transportation of personnel appears to be the leading cause[4] of fatalities followed by contact with equipment, fire and explosion, and exposure to harmful substances.

Table 1.1 – Major oil spill accidents		
Date	**Location**	**Details**
Apr 2010	Gulf of Mexico, USA	Spillage 4,900,000 barrels, 11 deaths due to explosion on Macondo exploration well.
Jul 2005	Mumbai, India	Fire in Mumbai High North platform after collision with vessel Samudra Suraksha killing 22 persons.
Mar 1992	Uzbekistan	Spillage 2,100,000 barrels from Mingbulak oil field.
Jan 1991	Persian Gulf, Iran	Estimated spillage 11,000,000 barrels resulting from Gulf War.
Jul 1988	North Sea, UK	Piper Alpha, 167 deaths
Oct 1983	Hainan Island, China	81 deaths due to sinking of drill ship Glomar Java.
Aug 1983	South Africa	Spillage 1,850,000 barrels from Castillo de Bellver (oil tanker).
Feb 1982	Newfoundland, Canada	84 deaths due to capsizing of Ocean Ranger drilling rig.
Mar 1980	Stavanger, Norway	Alexander Keilland rig sank, killing 123 workers
Jun 1979	Gulf of Mexico, USA	Spillage 3,500,000 barrels caused by blowout from Ixtoc 1 oil well.
Apr 1977	North Sea, UK	Spillage 1,900,000 barrels caused by blowout from Ekofisk oilfield.
Jan 1969	Santa Barbara, California, USA	Spillage 80,000 barrels

Table 1.2 – Major Helicopter Accidents[5]

Date	Location	Deaths/ Injuries	Details
May 2014	Takoradi, Ghana	4/4	Helicopter en route to Jack Ryan drill ship from Takoradi, crashed into sea.
Dec 2013	Bintulu, off Sarawak, Malaysia	0/8	When travelling offshore to land on a seismic survey vessel, helicopter ditched at sea.
Oct 2013	Venice, Louisiana, Gulf of Mexico, USA	1/2	4 persons on-board, 1 pilot died, 2 passengers with severe back injuries.
Jul 2013	Yatagun gas field, Andaman Sea, Off Myanmar	3/4	9 passengers and 2 crew.
Aug 2013	Off the coast of Shetland, North Sea, UK	4/14	16 passengers and 2 crew.
Apr 2009	Off Peterhead, UK	16/0	14 Passengers, 2 crew died.
March 2009	Offshore Newfoundland, Canada	17/1	18 persons on-board, 1 survived.
Feb 2009	East of Aberdeen, UK	0/18	16 passengers, 2 crew rescued
Nov 2006	Coast of Terennganu, Malaysia	1/20	21 persons on-board. 1 pilot died
Mar 1992	Near Cormant Alpha, NE of Shetland, UK	11/1	Super Puma AS332L
Nov 1986	Off Sumbrugh, UK	45/0	43 passengers and 2 crew died and only 2 persons survived in Boing Chinook crash which was returning from Brent Oil field.
Jul 1983	St Mary's, Isles of Scilly, UK	20/6	Helicopter crashed into sea killing 20 persons.
Aug 1981	Off Bacton, Norfolk, UK	13/0	Helicopter crashed into the sea killing all persons on-board.

1.2 Petroleum Formation and Exploration

Commonly known as 'fossil' fuels, oil and gas are produced from the remains of plants and animals buried millions of years ago under the surface of the Earth. As sedimentary layers formed over this organic material, the increased pressure and heat towards the centre of the Earth converted it to a complex organic matter known as 'kerogen'. When distilled, through porous rocks containing water, this matter produced petroleum products or hydrocarbons. These were forced upwards through porous rocks until they reached non-porous rocks, resulting in the formation of larger 'reservoirs' referred to as 'sedimentary basins'. When discovered, these basins containing oil or gas are called as 'oil or gas fields'. The main regions where these reservoirs have been discovered are the Gulf of Mexico, North Sea, Siberia, Middle East, North Africa, Indonesia, Campos/Santos Basins in Brazil, Newfoundland, Nova Scotia, Sakhalin in Russia, the east coast of India and the Caspian Sea.

On the basis of the stage a project is at, the petroleum industry divides itself into three sectors:

- Upstream: This sector includes the Exploration and Production (E&P) phase.
- Midstream: Once produced, the hydrocarbons are handed over to the midstream sector that processes, stores and transports the products.
- Downstream: This deals with the marketing of finishing products supplied from the midstream sector.

1.2.1 Petroleum Exploration

A typical offshore oil exploration and production project consists of five phases:

1. Exploration
2. Drilling
3. Completion
4. Production
5. Decommissioning.

Using the most modern technology including satellite imaging, the process of petroleum exploration begins by identifying areas of the Earth containing sedimentary basins. Once some evidence is obtained, seismic techniques are used to explore these basins. Sound waves are then used to study the underground rock formations likely to enclose reservoirs filled with gas or oil.

In areas of seabed known to contain petroleum products, specialised vessels are used to tow sonar navigation equipment that produces sound waves, also known as shock waves. Currently, one method available to generate shock waves is by using compressed air guns to shoot 'beams' of air into water (Figure 1.3). On land, other methods could be used such as thumping heavy plates into the ground, a method known as thumper trucking, or detonating explosives in holes drilled in land. Sometimes the latter method is also used over the seabed. The reflection of these waves varies with the type of rocks. Hydrophones (in water) or geophones (on land) are then used to receive echoes of sound (Figure 1.3). Patterns of these echoes are then analysed by geologists to determine the location of reservoirs. Computerised 3D seismic maps assist in pinpointing the most likely location for drilling for oil and gas. In addition, geologists also use gravity meters[6] and magnetometers to analyse gravity variations in the identified areas to see changes in the Earth's gravitational field confirming the flow of oil. If required, they also use 'chromatographs' to sniff hydrocarbons. Even then, the success rate for finding oil or gas is about 33 per cent which is a huge increase in the success rate from less than 10 per cent about 30 years ago.

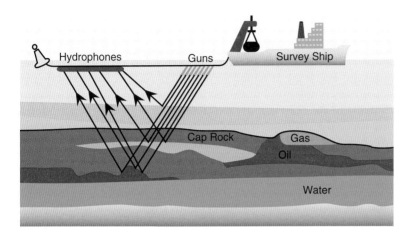

Figure 1.3 Petroleum exploration

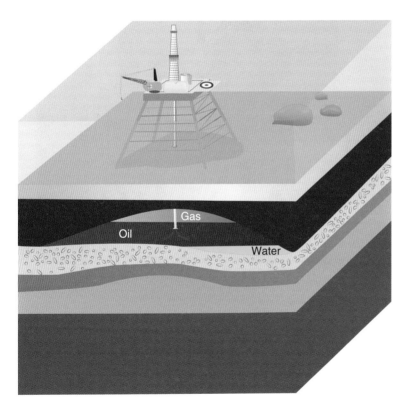

Figure 1.4 Drilling rig sketch

When seismic surveys gather concrete evidence about any seabed containing hydrocarbons, then necessary legislative procedures need to be followed to obtain leases, licences and consents. The process also involves assessing the impact to the local industry e.g. fishing as well as any impacts to the environment.

Figure 1.5 Offshore drilling rig (Source: Stena Drilling)

1.2.2 Drilling

During exploration stages, some drilling may be required to establish the exact location and size of the reservoir. Once established, larger scale drilling by using 'drilling rigs' (described in this chapter) is required to extract the petroleum products from the reservoir. The stage at which an oil well is readied for extraction is known as completion but the actual stage in which the petroleum contents of the reservoir are extracted is known as production. By this time, the drilling rigs are replaced by production platforms which extract petroleum products and distribute them to land based networks through pipelines or FPSOs (Floating Production, Storage and Offloading). When the contents of the reservoir are reduced, injection wells may used to inject water, CO_2 or steam to push the required contents out. Finally, the reservoir may become

Figure 1.6 A drill bit ready to be deployed

financially unviable when it has dried and hence may need to be abandoned. At this stage, the production rig may need to be decommissioned.

The hydrocarbons formed under the surface of the Earth, where either there is no 'non-porous' rock to enclose them or they are too close the surface, escape to the air. However, where these hydrocarbons are trapped several miles underground, geologists can locate them and then drilling is required to extract them.

Once the location of a reservoir is established, the first well that needs to be drilled is known as 'wildcat'. If hydrocarbons are discovered, the next step is to establish the size of the field to determine the quantities of discovered hydrocarbons. More wells known as 'appraisal wells' or 'step-out wells' are then drilled to gain further information. Depending upon the information obtained from the seismic surveys and size of reservoirs, the number of appraisal wells may vary from two to ten wells. Depth of water determines the method required for drilling. For example, in depths of 5 to 125 metres, jack-up rigs can be used whereas in up to 3500 metres depth, semisubmersible rigs or drill ships are likely to be used. Current methods of drilling are based on Edwin Drake's method for producing oil in which pipes were used to prevent boreholes collapsing thereby allowing deeper holes to be drilled.

1.2.2.1 Basics of Drilling

Whilst the basic processes for onshore and offshore drilling are similar, the challenges presented for offshore drilling are far more serious due to the remoteness of the location and harsher environments. Once the location of petroleum reservoirs under the seabed is established, a mechanical process to drill a wellbore through the seabed is carried out to extract this petroleum.

The industry uses three types of rigs to drill offshore wells. These are jack-up rigs, semisubmersible rigs and drill ships. Details of these rig types are discussed later in this chapter but the major components of a drilling rig are given below.

1.2.2.2 Wells

The types of well that can be found in the industry are primarily based on the type of product being extracted, for example, wells that can produce oil, a mixture of oil and gas, and wells that can produce only natural gas. Further classification of wells is based on the purpose they serve. For example:

- Wildcat wells – wells drilled in 'unknown' locations in search of petroleum reservoirs.
- Exploration wells – drilled specifically for finding new oil reservoirs.

- Appraisal wells – used to 'appraise' the characteristics of the contents of a reservoir.
- Production wells – specifically drilled for production of petroleum products.
- Abandoned wells – wells that have been plugged and therefore abandoned due to technical reasons.
- Suspended wells – a well on which operations are temporarily suspended because of various reasons but may be used at a later stage.

Each well is fitted with a 'wellhead' – a structure that acts as an interface between the drilling and production equipment and is placed on a casing head to seal the annular space. A 'Christmas tree' containing valves, spools, sensor, couplings, flanges and other fittings and connections is installed on each well. The flow of petroleum products is controlled through this tree. Generally, the offshore wellhead may be located on the production platform, known as a 'surface wellhead', or on the seabed, known as a 'sub-sea wellhead'.

1.2.2.3 Drilling Procedure

After drilling a several hundred metres deep shallow hole, metal tubing known as casing is forced into this hole. The usual diameter of this casing is 36 inches/91 cm. Connected with the drill pipe, the drill bit is then entered into the casing. As the drill bit cuts through the seabed, mud is pumped into the drill pipe. The cuttings of the rock formation are carried out of the drill pipe with the mud. When the hole is drilled to the required location, the drill bit and drill pipe are withdrawn to the drilling rig. Depending upon

Figure 1.7 Rig floor

the depth of the hole, a succession of smaller and stronger casings or liners are installed within each other as the bore continues deeper.

1.2.2.4 Rotary Drilling System

The rotary system uses rotation via a rotary table or top drive through which drill string and drill bit are rotated. The drill bit is the most important component used to penetrate through the Earth's crust, even the hardest rocks. The components used in a rotary drilling system can be divided into five groups:

1. **Prime movers:** Equipment that comes under prime movers is mainly related to the source of power for the drilling rig and all

Figure 1.8 Top drive system on drill floor

other components. This can be from the main power source for the rig or separate diesel generators.

2. **Hoisting equipment:** This is the most prominent structure on every rig and is used to hoist/lower the drill attachments and tools in or out of the well. Main components include mechanical winch (drawworks), derrick, crown block, travelling block, hook and wire rope. For the comparatively small but most important task of replacing a worn drill bit or changing it for different types of rock, the whole drill string needs to be hoisted using the hoisting equipment. Details of various components of hoisting equipment are given below.

   a. **Mechanical Winch or Drawworks:** Commonly referred to as drawworks in the oil and gas industry, the winch is used to hoist or lower the drill string in the borehole. The drill line is stowed on a drum which is attached to a hydraulic or electric motor. Four prominent features of the drawworks are the braking, transmission (gear) systems and a drive sprocket to rotate the rotary table. Many rigs use a standalone drive for the rotary table or top drive. The braking and transmission systems provide the driller with thorough control of the operations. The fourth feature consists of two catheads which are used to spin and tighten the drill pipe joints whereas the second cathead is used to loosen the drillpipe, hence the names make-up cathead and breakout cathead respectively.

   b. **Derrick:** Also known as 'mast', this commonly consists of a pyramid structure supported by four legs placed on a square base. The derrick and other rotating equipment is located on the rig floor which also provides work space for drilling operations. The height of the derrick determines the

Figure 1.9 Drill bit with its nozzles

length of the sections of drill pipe, hence depending upon the requirements, the derrick height may be varied. All load is supported by the derrick at any time when the drill stem is attached to the travelling block and drill line.

3. **Rotating equipment:** The components that rotate the drill bit form part of this equipment. These include:
   - **Swivel:** This is attached to the drilling hook and allows all the attachments below it to rotate freely. Its design ensures a pressure tight seal on the hole whilst a gooseneck connection on the swivel allows the drilling fluid to be introduced into the drill stem. For a top drive system, a power swivel is used to rotate the drill string instead of a rotary table.
   - **Rotary table/top drive:** The rotary table transfers rotational force to the drill via the drill string. The top drive serves the same purpose but is located at the swivel. The benefit is the ability to use longer sections of drill pipe in comparison with the rotary table which can only use sections of about 9 m length. The top drive can use about 27 m long drill pipes joined together in one length. As a result, there is considerable time saving during drilling operations.
   - **Drill string:** This consists of tubulars inserted in the hole for drilling and consists of the drill pipe, Heavy Weight Drill Pipe (HWDP), drill collars, subs, stabilisers, float valve and jars. The lowermost part of the drill string is called the Bottom Hole Assembly (BHA) which consists of the drill bit, mud motor, stabilisers and subs.
   - **Drill bit:** Attached to the drill string, this is the end of the drill used to cut through rocks. Built from extremely hard materials such as tungsten carbide steel or diamond, its shape may vary.

Figure 1.10 Shale shaker

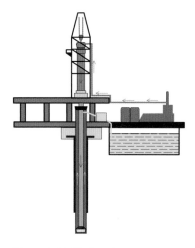

Figure 1.11 Mud system

Figure 1.12 BOP

- **Drill collars:** These are heavyweight tubulars used at the bottom of the BHA to provide weight on the drill bit and also to prevent buckling of the drill pipe. These can come in a flush or spiral surface. Drill pipes and drill collars have threaded connections known as tool joints.
- **Kelly:** This is a square or hexagonal pipe of approximately 12 metres length, attached via the kelly bushing to the rotary table through the master bushing. Gearing in the rotary table rotates the master and kelly bushing thus rotating everything below it.
- **Stabilisers:** They are used to keep the drill string in the centre of the borehole and therefore reduce the area of contact between the components with the borehole.
- **Jar:** A tool used to free components stuck in the drill string.
- **Mud motor:** Also known as the drilling motor, this is a pump placed in the drill string to provide additional power to the drill bit during drilling. Power is provided by a positive displacement motor driven by mud (drilling fluid) circulation. For directional drilling, a motor is used to alter the direction to the required angle and steer the drill bit to its target location.
- **Mud:** This is the most common drilling fluid made from a mixture of clay, chemical, water, oil and other weighting materials.
- **Subs:** Any short length of pipe, collar or casing used in the system.
4. **Circulating equipment:** This equipment is used to pump the drilling fluid from surface tanks, through the drill string and back to the surface. The components of the system also ensure cooling and lubricating the drill bit, controlling well pressure, removing the debris e.g. rock cuttings and ensuring walls of the well are lined with mud. The main components of this system are:
    - **Pump:** This pumps the mud to the drilling equipment after sucking it from mud pits. Mud is returned to the mud pit via the mud-return line.
    - **Mud pits:** A tank to mix and recycle drilling mud. New mud is mixed in the mud-mixing hopper.
    - **Shale shaker:** This takes out rock cuttings from drilling mud and puts them into a reserve pit via the shale slide.
5. **Blowout preventer** (BOP): Accidental and therefore uncontrolled flow of gas or oil from the reservoir is known as blowout (or uncontrollable kick). The BOP is one or more high pressure valves fitted on the seabed/at the top of well on high pressure drill lines to perform a controlled relief of pressure so that a blowout can be prevented. The first line of defence for prevention of a blowout is mud. If this is not successful, then the BOP comes into use.

1.2.2.5 Directional Drilling

Production wells may not be located directly underneath the production platform. When multiple wells need to be drilled through the same vertical wellbore, a technique known as deviated or directional drilling is used. There are a number of benefits of using this technique including an evenly spread extraction around a central location as well as minimising the impact of drilling on the environment. If a petroleum reservoir is discovered through this process, a number of sub-sea wellheads can be grouped together and connected to the main platform.

In many drilling projects, the wellbore is deviated from the vertical direction to a planned path towards a known reservoir containing hydrocarbons. This technique has been used in the industry since the 1920s and allows for multiple wells from the same vertical wellbore. The benefits of this technique include reduced environmental damage and significant savings in time and therefore cost of drilling wellbores. Additionally, many sites may not provide an easy access to position a drilling rig due to the hydrocarbons being located in harbours or estuaries, shipping lanes or under mountains. Occasionally, the wellbore has to be deviated from its originally planned path due to, for example, a stuck drill string. In such case, the directional drilling is known as 'side tracking'.

With advancement in technology, directional drilling now utilises Global Positioning Systems (GPS) for positional accuracy, advanced survey techniques for gathering the most accurate geological information, 3D measuring devices and specialised drill bits.

1.2.2.6 Personnel on Drilling Rigs

The roles described here may vary from one rig to another depending upon the size of the rig, nature of operations, the stage a project is at e.g. exploration or production stage. Complying with company policies and ensuring own and others' health and safety are generic requirements for each role. Other main responsibilities for main roles are given below.

1. **Offshore Installation Manager (OIM):** The OIM has overall responsibility for the safety and security of an offshore installation and successful completion of all operations in the same way as the captain of a ship. Generally distinction is made between a rig manager (described in the next section) who is in charge of a drilling rig and an OIM who is in charge of an offshore production platform. The OIM's duties generally include:
   - ensuring compliance with company procedures and regulatory/legal requirements;
   - managing emergencies on-board;

- ensuring that the rig manager and operator's representative are kept informed about all activities;
- liaising with contractors to ensure agreed plans are executed safely;
- developing, managing and participating in emergency response drills and preparing reports for any such activities;
- conducting accident investigations and preparing reports;
- reporting on work progress against agreed targets to shore based management.

2. **Rig Manager:** The rig manager has overall responsibility for the conduct of safe operations on drill ships where the master may also be a rig manager or there may be a separate rig manager. A summary of requirements for this role is given below:
   - ensuring full compliance with national/international and company requirements for navigation, maintenance and operations;
   - planning and executing the planned drilling schedule as agreed with the client representative and other heads of departments;
   - managing and then supervising after delegation of drilling, marine, catering and maintenance roles to the respective heads of departments;
   - preparing and managing budgets for overall operations of the rig;
   - ensuring an up-to-date inventory and requisition is available for ordering required supplies;
   - working with client representatives to agree and/or amend any plans;
   - ensuring effective and safe planning for rig moves;
   - ensuring training and career development of relevant staff;
   - ensuring safety of health, environment and the rig and third party vessels by taking all necessary actions.

3. **Barge Master:** Usually applicable to drill ships in the offshore oil and gas industry, a barge master is in overall charge of navigation, safety and security of all operations of the vessel. For a fixed installation, the OIM may be assigned these tasks. Slight variations may be found in many cases where a barge master reports to the OIM instead of the rig manager.

4. **Maintenance Supervisor:** As the title indicates, the maintenance supervisor is in overall charge of maintenance of electrical and mechanical systems and structures on-board a rig. Other duties include:
   - planning, delegating and supervising mechanical and electrical work for drilling operations;
   - supervising the work of electricians, mechanics and motormen;

- liaising with shore based engineers or suppliers for any maintenance or task issues;
- ensuring health and safety of all employees and effective implementation of permit to work system.

5. **Assistant Driller, Driller, Drilling Superintendent:** These are three ranks based on qualifications and experience. A driller supervises the drilling personnel and controls the rate at which drilling takes place. Duties include:
   - maintaining control of the drill floor;
   - monitoring accurate connection of the drilling pipes and proper use of drilling tools;
   - operating the machinery to raise or lower the drill;
   - record keeping for all drilling operations;
   - keeping a check on the BOP and if blowout occurs, handling it effectively;
   - carrying out preventative maintenance and repair of the equipment used;
   - carrying out all drilling operations as per the agreed drilling plan;
   - ensuring all other personnel such as derrickhand, roustabouts, floorhand, etc. are fully informed of work expectations;
   - ensuring health and safety of all workers.

6. **Crane Operator:** Operates cranes/derricks as required and keeping in line with the following duties in general:
   - lifting operations for casings, pipes, containers, machinery, equipment or other supplies for use on the rig;
   - lifting supplies to/from the support vessels such as crew boats, barges or other vessels;
   - operating personnel lifting baskets with the cranes;
   - supervising roustabout and other personnel as required;
   - assisting welders, electricians or mechanics as required.

7. **Derrickhand:** The main job is to assist the driller in the deployment or withdrawal of the drill pipe, casing or tubulars through the rotary table. Other general duties include:
   - carrying out preventative maintenance of the derrick/lifting equipment, its accessories and other assigned equipment;
   - ensuring safe and effective operation of the mud equipment and pumps;
   - assisting the driller and floorhands on the drill floor;
   - supervising the roustabouts and floorhands as required during operations.

8. **Chief Electrician, Senior Electrician and Electrician:** Again, with various ranks based on experience, the electrician is responsible for the efficient provision of all required electric equipment for the rig. Examples of duties include:
   - carrying out preventative maintenance on all electrical equipment or components;

- carrying out routine checks to ensure the efficient running of all electrical systems and machinery including refrigeration, air conditioning, cooling, power generation/distribution and control systems.

9. **Floorhand:** The floorhand is employed to handle tubulars and drilling tools on the rig floor when working with the drill string. Other duties include:
   - assisting the driller as directed;
   - assisting in connecting/disconnecting drill pipes, tubulars, casings, etc.;
   - assisting the derrickhand for mud related operations and relieving them when fully trained; carrying out maintenance and upkeep of equipment on drill floor.

10. **Chief Mechanic and Mechanic:** Examples of main duties include:
    - troubleshooting and resolving problems resulting from any breakdown of equipment or machinery;
    - carrying out preventative maintenance of all assigned equipment;
    - supervising and training motormen;
    - assisting electrician in carrying out repairs to motors and cooling, refrigeration, air conditioning systems, etc.

11. **Motorman:** The main job is to assist the mechanic. Other duties include:
    - carrying out preventative maintenance of mechanical equipment including engines, pumps, water/steam lines and boilers;
    - general housekeeping duties for machinery;
    - assisting in movement of equipment and machinery on-board the rig.

12. **Toolpusher:** One of the senior roles on the rig. Some rigs also employ a senior toolpusher. Main duties include:
    - overall responsibility as head of drilling department for all operations and reporting to either the OIM or master;
    - ensuring all supplies including equipment, spare parts and tools are available;
    - may also be required to perform administrative tasks such as rig crew payroll, overtime and other matters related to personnel.

13. **Roustabout:** This role involves basic labouring tasks to maintain the workplace in good order. On some rigs, there may be a lead roustabout to supervise other roustabouts. General duties of a roustabout include:
    - cleaning and painting decks, machinery and work equipment;
    - ensuring equipment is supplied or removed from the work site;
    - operating crane/derrick/winch;
    - assisting in repairing equipment;

- working on the circulating equipment, particularly mud-mixing equipment.

14. **Roughneck:** Experienced roustabouts can be promoted to roughneck. As a result, the duties will require more skill. Examples of duties include:
   - connecting drill pipes as the drill hole becomes deeper;
   - maintenance and upkeep of the drilling equipment;
   - operating crane/derrick/winch;
   - extracting or inserting drill.

15. **Safety Representative:** This may be an additional role assigned to any of the existing workers to represent workforce safety concerns to the safety officers and attend safety committee meetings. They may be required to assist the safety officer.

16. **Safety Officer:** The main job is to assist the rig based management in ensuring compliance with health, safety and environment policies and legislative requirements. Additional jobs include carrying out audits, incident/accident investigations and training with specific reference to health and safety on-board. The main duties of a safety officer are:
   - organising rig and company induction programme for new and existing employees;
   - developing health and safety policies and procedures for the rig;
   - ensuring compliance with company, operator and regulatory authority/legislative requirements;
   - organising, coordinating external training and if required delivering training programmes to the rig crew;
   - assisting with investigation of incidents and accidents and making recommendations to prevent recurrence;
   - offering safety briefings to the rig personnel based on the findings of accident investigations;
   - implementing company 'safety observation systems'.

17. **Welder:** The main job is to cut, shape and join metals to repair or fabricate structures or machinery.

18. **Operator's Representative:** The role is to ensure all work is carried out as required by the rig charterer or the well operator. Other personnel such as surveyors or geologists may also be employed in addition to the operator's representative.

19. **Mud Engineer:** Also known as drilling fluids engineer or simply 'mudman', the main duties include:
   - ensuring operations requiring drilling and completion fluids are completed effectively in line with company procedures;
   - liaising with 'mud specialists' for problem solving, mud/fluid parameters and agreeing changes in specifications;
   - advising on problems related to drilling and recommending any alternative solutions;

- providing technical guidance for drilling fluids;
- ensuring an adequate stock is available on site and keeping a record of material usage;
- ensuring mud plant is effectively maintained and operated.

20. **Sub-sea Engineer:** In conjunction with the sub-sea supervisor, the main job is to ensure effective operation of the BOP and control systems. In addition, the sub-sea engineer also supervises installation and retrieval of BOPs. Other duties include:
- providing technical guidance for operation and maintenance of sub-sea equipment;
- overseeing operational procedures and carrying out audits to ensure their suitability;
- developing and implementing the equipment maintenance schedule.

21. **Cementer:** The main job is to prepare a well for further drilling by pumping cement as required to support the wellbore. Other main duties include:
- performing calculations for the proper mixture of cement slurry required to be pumped into the wellbore;
- where required, ensures carrying out pressure testing of the equipment.

22. **Offshore Medic:** The main job is to ensure the health, safety and welfare of the rig crews. Duties include:
- carrying out medical examinations of the crew members;
- providing first aid and dealing with injuries or illnesses on-board and administering drugs if required;
- maintaining stock of medical equipment and medicines;
- organising external medical assistance or evacuations for shore based treatment;
- keeping the OIM/Master informed of any medical issues on-board the rig;
- after suitable training, carrying out other administrative or operational tasks as required. An example of such role is helicopter landing officer.

Other variants or additional roles may be found on various rigs depending upon the size and nature of operations and type of equipment used. Example of such roles include:

- **Electro Technical Officer/Engineer:** tests, maintains and repairs the electronic equipment.
- **SAP planner:** the SAP (Systems Application Products) planner looks after the maintenance management system, updates records for maintenance and highlights any equipment requiring maintenance. The job also includes ensuring sufficient inventory of spare parts. The most common system used for this job is P3M (SAP) maintenance management system.

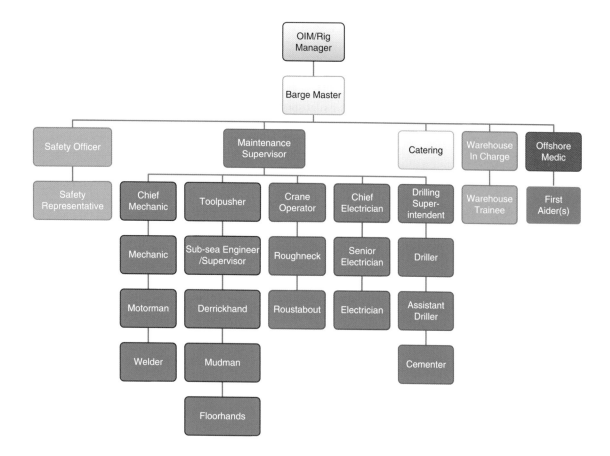

Figure 1.13 Typical drilling rig organogram

- **Dynamic Positioning Officer (DPO):** qualified to officer of the watch, chief mate or master level, the DPO operates the automatic positioning equipment to maintain the position of the vessel.
- **Barge Engineer:** usually a marine engineer by qualification, his main job is to carry out operations and maintenance of the marine equipment on the vessel. Other additional duties may also be assigned to this role.
- **Logistics Coordinator:** the main task for this role is to ensure sufficient inventory of spare parts and materials in conjunction with the SAP planner. In addition, the logistics coordinator is also required to keep a record of all consumables.

1.3 Offshore Drilling and Production Rigs

As petroleum exploration reached further offshore, deep water drilling required sophisticated equipment including drilling rigs to drill in water depths of over 3000 m. Consequently, the offshore structures including drilling and production rigs, collectively referred to here as rigs, had to be developed to ensure they could meet the demands and withstand the forces due to:

1. dead or permanent loads e.g. weight of the structures, consumables, ballast, additional equipment and underwater structures as well as hydrostatic pressure on underwater structure;
2. operational loads due to non-permanent equipment and materials, e.g. vibration due to operational machinery, forces due to lifting weights and installation of various components and helicopter operations, etc.;
3. environmental loads caused by wind on structures above water, waves on below water structures and other factors such as earthquakes, etc.; additional factors include the impact of ice or snow, temperature variations and marine growth;
4. accidental loads that may arise as a result of exceptional circumstances such as the impact of collision with another vessel, fire or explosion and flooding.

The above loads are taken into consideration during the design of each offshore structure. Under normal operating circumstances, the design incorporates operational and environmental loads; whereas for extreme operating circumstances, extreme conditions are considered. Generally, each structure design must meet three criteria – operations, survival and transit i.e. the maximum forces a structure is designed to withstand during its operations, extreme conditions and when it is being transported to the installation location. In addition, various damage conditions are also considered during the design phase to assess the survivability of the structure in extreme circumstances. In order to achieve the required structural strengths, the industry complies with the offshore standards agreed within the industry such as those from the classifications societies or for a specific country such as the British 'Code of Practice for Fixed Offshore Structures'.

1.3.1 Mobile Offshore Drilling Rigs

Mobile offshore drilling rigs are generally referred to as Mobile Offshore Drilling Units or MODUs. Their design life is considered to be 25 years when deployed in the offshore environment and 50–75 years when deployed in the inland waterways, lakes or rivers.

There are a number of types of MODUs but commonly the industry divides them into two types i.e. floating rigs and bottom supported rigs. The type of MODU used in any area will depend upon the depth of water, distance from shore and prevailing weather and other factors affecting the operations of the MODU. Design consideration therefore takes into account the loads that may be exerted by extreme winds and waves combined with motion

Figure 1.14 BP's Deep Water Spar Platform Drill Rig 'Mad Dog' (Source: BP plc)

of the structure. Various types of drilling/production rigs and their components are described in the next subsections.

1.3.2 Topsides

In an oil rig, the facilities used for production or processing of petroleum products are often referred to as 'topsides'. The construction of topsides may be modular with each module weighing up to many hundreds of tons. These are fabricated onshore and then transported and installed onto the base structure which may be any one of the construction types discussed in the next sections. Heavy lift cranes are used to place these modules on top of the base structure. A large number of personnel are involved in the stage of construction known as 'hooking up' which requires large accommodation provided in the form of 'floatels'. In some cases, floatels are removed after completion of this stage but in many cases, floatels are kept in place to provide accommodation for rig personnel. Usually floatels are connected to the main rig by means of telescopic gangways.

The base structure supports the topsides above the water level and consists of the following components:

- facilities to separate oil, gas, water, sand including measurement and treatment facilities for disposal of any wastes;
- drilling facilities;
- sewage, water, electricity, fuel and other utilities;
- cranes or derricks;
- accommodation for rig crews;
- workshops;

Figure 1.15 Topside module being fitted with heavy lift crane (Source: BP plc)

Figure 1.16 Types of offshore structures (from left to right: 1. Fixed platform; 2. Concrete gravity platform; 3. FPSO/drill ship; 4. Normally unmanned installation (NUI); 5. Steel jacket platform/compliant tower; 6. Floating production platform/ tension leg platform; 7. Fixed platform; 8. Fixed platform)

- control room;
- safety systems e.g. fire detection systems, deluge systems;
- helideck – generally fitted in such a way that the helicopter rotor blades can be kept clear of any projected structures from the rig;
- flare boom.

1.3.3 Bottom Supported Rigs

1.3.3.1 Fixed Platform

These structures are attached to the seabed with concrete or steel legs which support the deck containing the drilling rig, production equipment and accommodation. Attachment to the seabed is either by means of piles or with the weight of the concrete legs. This type of structure is used in areas where larger petroleum reservoirs are discovered, the structure is required for long-term use and depths are generally around 500 m. Four sub-types of structures can be used to build the legs, i.e. steel jackets, concrete based gravity structures, floating steel or floating concrete.

1. **Steel Jackets:** these structures consist of a jacket, piles and decks, commonly used in the Persian Gulf, Nigeria, Mexico and California. The jacket is fabricated from steel welded pipes and attached to the seabed by using steel piles. The piles can be up to 2 m in diameter and inserted into the seabed for depths of up to 100 m. Legs attached to the piles support the weight of the top structure.
2. **Concrete Caisson and Concrete Gravity Based Structures (CGBS):** a caisson is a watertight concrete structure kept on the seabed to provide a base for CGBS.[7] The compartments within a caisson can be used for oil processing or storage. These

compartments also act as buoyancy tanks but were first deployed in the North Sea in the 1970s for large oil fields requiring large processing facilities. The first concrete platform was installed in Ekofisk field in the North Sea in 1973. Due to immature pipeline infrastructure to export oil to land, CGBS provided storage facilities for oil from where it was loaded in tankers.

Their design varies considerably with weights ranging from 3000 to 1.2 million tonnes providing reliable storage facilities that can withstand extreme environmental impacts. A CGBS consists of a concrete base or caisson with a number of piles to support the deck on which the top platform is located. The type of seabed determines the size of base required. There are generally three types of CGBS i.e.

a. with a single caisson extending up to 16 m above sea level; the caisson can have a diameter of up to 50 m and supports the topside structure;
b. with up to four concrete pile shafts extending above sea level;
c. with steel jack-up legs supporting the topside structure; the base caisson can be used for storage of oil.

3. **Floating Concrete:** Concrete has been used in marine structures since the early 1900s. Floating concrete structures are held in position by anchored chains or wires. Generally these are used for production or storage units due to their ability to move within the slack provided by the anchor chains or wires.

Figure 1.17 Jack-up rig with production platform shown in yellow

1.3.3.2 Jack-up Rigs

Jack-up rigs can be compared to 'drilling barges' except that when these are brought to the required position, their legs, generally three or four, are lowered to the seabed, as a result of which the jack-up rig remains above the surface of water. Since the full support to the 'topside' is provided by the legs, the strength of the seabed on which these legs are resting needs to be carefully assessed. As long as the legs are supporting the main structure, the rig can remain literally fixed at one location. Their deployment in water depths of over 150 metres is considered unsafe due to larger lateral forces resulting from the weight of structure and movement of water. Additional strengthening of the legs will make them economically unviable.

Drilling is carried out either through a moonpool or a drilling rig fitted on cantilever beams on a skidding system. The latter method allows the drilling unit to be cantilevered from its storage position to the operational or drilling position. This enables the drill floor structures to move both longitudinally and transversely on the deck of the jack-up rig where the drill operation end of the cantilever extends beyond the platform.

1.3.3.3 Compliant Tower

The compliant tower is similar to the fixed platform in the way they are supported by tubular jackets to support the topsides. They are attached to the seabed by using piles and are the tallest free standing structures with a small seabed footprint. However, they withstand the forces of wind and water in the same way as floating structures because of the larger depths of around 1000 metres in which they are installed. In addition, the upper part of the structure may additionally be supported by mooring chains or wires to the seabed, in which case they are referred to as 'guyed towers'. A number of buoyant tanks are located in the upper part of the jacket. The amount of buoyancy in these tanks is controlled to ensure the required tension for the prevailing weather conditions. Generally, the size of the topsides on the compliant towers is smaller than the fixed platforms due to smaller supporting jacket structures.

1.3.3.4 Tension Leg Platform (TLP)

This type of structure consists of a buoyant hull comprising a number of vertical cylindrical columns and horizontal submerged pontoons braced by tubulars. The TLP is held in position by a mooring system which consists of tension legs attached to the platform and connected to foundations on the seabed. The foundations can be anchors, gravity foundations or more commonly piles driven into the seabed. If piles are used, their scantlings and materials depend upon the depth of water and expected environmental conditions but can be around 100 m in length. Generally, steel or concrete piles are used in the industry. Tendons are then used to connect the foundation to the platform.

The buoyancy of the hull maintains tension in the moorings enabling the rig to remain in position. This system allows for horizontal movement but does not permit vertical movement. There are three different types of TLPs depending upon their deployment depths and weather factors. These are full size, mini TLPs and well-head TLPs.

1.3.4 Floating Rigs

1.3.4.1 Drill Barge

Drilling barges are used for shallow water inland areas such as rivers, canals and lakes, etc. Since the drilling barges do not have their own propulsion systems, they are towed to the location by other vessels.

1.3.4.2 Drill Ship

A drill ship is a ship fitted with drilling equipment which has the ability to drill in deep waters. All drill ships are fitted with a 'moonpool' or 'drill slot' in the midships region. This is an opening in the hull through which equipment can be deployed. However, because of their ship shaped hulls, they are more susceptible to wind and wave action which limits their deployment to moderate climatic conditions. Their benefits include their ability to operate in the areas of the sea in which they can be utilised. On the other hand, their self-propulsion at faster speeds, larger storage capacities for consumables and drilling products reduce the logistical complications further offshore. They are also fitted with a helideck at the forward end of the vessel and a heavy crane for handling

Figure 1.18 Drill ship

equipment. Depending upon the depth of water and other variables such as wind and current, etc., drill ships can maintain position by either the use of mooring or DP (Dynamic Positioning) systems or a combination thereof. With the advancement in technology, drill ships can easily drill in depths of up to 1000 metres.

1.3.4.3 Semisubmersible

Generally these are divided into two types: inshore and offshore submersible rigs. Inshore rigs cannot be deployed offshore because their structures are extremely vulnerable to offshore impacts of wind and waves.

The offshore submersible rigs consist of a topside structure mounted on pontoons via columns. When these columns are flooded, they cause the pontoons to submerge to the required depth under water. These types of rigs are usually deployed in depths of 600–3000 m, typically in South America and the Gulf of Mexico. In some cases, DP systems are used to maintain their position but mooring systems may also be used. Semisubmersible rigs are very common in the industry due to their better stability characteristics.

There are two main types of semisubmersible rigs: column stabilised and bottle type. The column stabilised rig appears to be more common in the industry. The topside is connected by vertical columns to the two horizontal structures. The bottle type semisubmersible rig consists of a bottle shaped hull below the topside structure which is submerged under water. These rigs can be towed to the required location.

The predecessors of the semisubmersibles were submersible rigs which were similar in construction to the former. However, their columns were filled with ballast water to set them on the seabed. As a result, the depth in which these could operate was only between 20–30 metres.

1.3.4.4 Tender Rig

As the existing platform structures deteriorate with time, new wells may need to be drilled in the vicinity of existing ones. In such situations, a technique referred to as 'Tender Assisted Drilling' or TAD[8] may be used. Generally fitted with a large crane, open decks and crew accommodation facilities, they are easy to deploy and therefore economical.

The tender has the ability to transport the Drilling Equipment Set (DES) to the desired location. When in position, the tender will be moored close to the location to drill a well. The drilling will be carried out directly from the production platform or a drilling barge may be deployed. The DES may be transferred by the tender rig as required.

1.3.4.5 Floating Production Systems (FPS)

A floating production system is fitted with equipment found on fixed platforms to extract oil from sub-sea wells located in moderate depths. They generally have semisubmersible hulls for purpose built FPSs but there may be ship-shaped mono-hulls converted from oil tankers. A mooring system consisting of a spread of mooring chains or wires is used to keep the FPS in position. A mono-hull provides an additional benefit of using the tanks for storage, effectively making the FPS an FPSO.

1.3.4.6 Floating Production Storage and Offloading (FPSO)

The main purpose of these vessels, as the name dictates, is to process petroleum products, store and offload them into other shuttle tankers or shore based facilities through submarine pipelines. These vessels can be purpose built or converted from an existing oil tanker. Mainly used for liquid petroleum products, the industry is now diverting towards an similar concept for floating liquefied natural and petroleum gases as well.

These vessels are particularly effective in deep water locations where pipeline transport of petroleum products is not economically feasible, especially where the reservoirs are not of a very large size. A further benefit is that an FPSO can easily be relocated when the field becomes depleted of the petroleum products or in cases of extreme weather due to storms, cyclones or icebergs, etc.

FPSOs are usually fitted with production equipment usually found on a production platform. This equipment consists of oil processing, gas treatment, water separation/injection and gas

Figure 1.19 FPSO in operation
(Source: BP plc)

compression systems. Where an FPSO is not fitted with oil and gas production or processing facilities, it becomes an FSO (Floating Storage and Offloading) which seem to have been used in many areas around the world. A typical feature of FPSOs is their mooring system fitted to the FPSO hull. This system allows the vessel to rotate freely as dictated by weather conditions. The main types of FPSO mooring systems are:

1. **Spread Mooring System:** This system allows an FPSO to moor at a fixed location on a fixed heading. Generally, four groups of anchor legs are used in a symmetrical pattern on the bow and stern of the vessel. This arrangement can be used on any vessel size in any depth.

2. **Single Point Mooring (SPM):** An SPM consists of three major parts: an anchoring system, the main body of the buoy and a petroleum product transfer system. A single point mooring provides a link between the oil production/storage facility and a vessel loading/offloading this oil. An SPM is necessarily a floating

Figure 1.20 Riser with heave compensation (Source: BP plc)

buoy anchored offshore to allow this operation. Generally located many miles away from shore, this facility can provide economical means of loading vessels without any complications of bringing them into a harbour. Anchors, anchor chains or wires are used to moor the SPM to the seabed. The mooring system used allows several degrees of freedom of movement within defined limits to withstand the effect of current, wind, waves and loading conditions of the vessel. The body of the SPM contains a 'rotating' table that connects to the vessel allowing the vessel to rotate around the SPM. The petroleum product transfer system known as a riser uses flexible hoses which connect the sub-sea pipelines to the SPM's transfer system through a swivel system. Further variations in the SPM technology are described below.

a. **Turret Mooring System:** This can be:
   i. Internal turret i.e. fitted within the vessel's forward end. The turret arrangement consists of a large roller bearing fitted inside a moonpool. The major benefit is that the vessel can rotate freely under the effect of external forces of wind, current and tide, etc. An extra benefit is the ability of the vessel to detach from this system to take shelter in extreme weather conditions such as storms, etc. However, the initial cost of a turret system is extremely high in comparison with spread mooring systems.
   ii. External turret: This arrangement is similar to the internal turret but it is fitted in a steel structure attached to the bow or stern of the vessel. The mooring chains and oil/gas transfer lines are attached to a fixed chaintable. The benefits of this system are similar to the internal turret but the costs are comparatively less.

b. **CALM (Catenary Anchor Leg Mooring):** CALM primarily consists of a floating buoy secured to the seabed by using catenary chains attached to anchors or piles. The buoy is fitted with a turntable on top of the buoy where mooring rope/chain from an FPSO or an oil tanker can be attached. The petroleum products can be transferred to the vessel through sub-sea hose systems. Generally used in depths of up to 100 m, CALM provides free rotational movement to the vessel moored to this buoy used for both importing and exporting oil products.

c. **SALM (Single Anchor Leg Mooring):** Similar to CALM, SALM is a facility mainly used for unloading offshore petroleum products. In order to eliminate the damage caused to the loading pipe by coming in contact with mooring chains, a single chain mooring leg is used instead of the catenary. The chain is connected to the seabed either by piles or by using weighted ballast.

3. **Dynamic Positioning (DP):** A DP system is defined[9] as one that is able to reliably keep a vessel in position when working up

to the rated environment, such that the maximum excursion from vessel motions (surge, sway and yaw) and position control system accuracy (DP footprint) is equal to, or less than, half the critical excursion for the work being carried out.

For almost all offshore floating vessel operations, the effect of wind and waves can create several problems in maintaining position. For example, a drill ship will be unable to carry out drilling operations if its position were not 'fixed' at one place. Consequently, systems are required to enable offshore vessels to maintain their position and withstand the impact of wind, waves and currents. Such ability is provided by employing DP systems which consist of a number of thruster units or propellers controlled by a computerised system connected to positioning sensors on the seabed in addition to obtaining positioning information from GPS systems. Further information about prevailing weather is also fed to the computers which automatically engage thrusters to maintain the vessel's position.

1.3.4.7 Floatels

Offshore accommodation vessels, commonly known as 'floatels', provide additional accommodation and construction support for offshore projects such as:

- assisting drilling rigs;
- commissioning of new fields;
- support for existing rigs;
- decommissioning.

Facilities in floatels include:

- accommodation for a large number of personnel;
- welfare facilities;
- catering facilities;
- storage for equipment, fresh water, fuel oils and sewage;
- workshops;
- offices;
- medical facilities;
- cranes;
- mooring and towing equipment.

Many of these vessels are self-propelled and fitted with DP systems. However, there are other offshore rig types such as 'semisubmersibles' that operate on the same principle as other offshore rigs. The vessels that are purely designed to support offshore projects and provide accommodation are also known as 'Accommodation Support Vessels' or ASVs.

1.3.5 Sub-sea Production Systems

Most offshore developments consist of multiple wells, requiring these wells to use a single production system to control costs. The system therefore incorporates sub-sea separation, re-injection or boosting facilities.

The salient feature of this system is the location of production equipment on the seabed instead of the floating rig. Consequently, unwanted components that would otherwise be processed on-board the rig are processed on the seabed reducing the requirements for risers, flow lines and associated fitments and therefore significantly reducing the costs as well as making the system more efficient. For example, higher level sub-sea boosting is required in deep water fields. By using sub-sea production systems, the produced gas, water and waste can be re-injected to the reservoir thereby increasing the pressure and easing the production. Furthermore, the unwanted waste can also be re-injected.

The first system of this type was installed in the Tordis field in the North Sea in 2007. Shell's BC-10 project in Brazil began production in 2009 and consists of six sub-sea gas–liquid separators and booster combined with an FPSO for storage.

Figure 1.21 Semisubmersible accommodation vessel (Source: © Prosafe)

1.4 Specialised Vessels

1.4.1 Seismic Survey Vessels

Seismic survey vessels are specialised ships used to carry out seabed surveys in high seas and oceans to locate hydrocarbon reservoirs.

These vessels utilise seismic waves transmitted towards a pre-defined areas of the seabed. The returned signals are then analysed on-board the vessel by using specialised equipment. The results obtained indicate whether the seabed contains any hidden reservoirs.

In this technique, the two components of the returned signals include the strength of reflected waves and the time it takes to travel through various layers of earth in the seabed and back to the sensors. The data obtained can be translated into 2D or 3D graphics.

The sensors used are known as 'hydrophones' capable of detecting pressure fluctuations in the seismic waves returned from the seabed. The hydrophones are placed along a cable called a streamer. A number of streamers may be towed behind a survey vessel and can be up to 12 kilometres in length. These streamers are spread by diverters that extend the coverage area to around

Figure 1.22 Seismic survey vessel

500 metres. The end of each streamer is identified by a tail-buoy fitted with a radar reflector and flashing lights to show its position for other traffic in the vicinity of the vessel. Additional boats may be utilised to assist the towage of streamers and to warn other vessels operating in the vicinity of streamers. The speed at which streamers are towed is usually 4–5 knots to enable complete profiling of the seabed.

These vessels are fitted (a projected structure on the bottom of the vessel) with blister and moonpool for seismic instruments, storage reel cradles, towing rope along with winches for streamer storage, etc., specialised winches for streamers, helideck and satellite communication and position fixing systems. Additional functions of these vessels include the study of the geology of the seabed in the areas under investigation. The information provides data about the formation of rocks, trenches and other materials in the seabed. Once initial data indicates the presence of hydrocarbons, further surveys are carried out to determine the exact nature of the seabed to pinpoint the exact drilling location.

1.4.2 Anchor Handling Tug Supply Vessels

Anchor handling tug supply vessels (AHTS) are mainly used to tow and then anchor an oil rig to its desired location, with their main job being the deployment or recovery of the sub-sea mooring system for offshore structures. Depending upon the nature of the job, one or more AHTS can be deployed at the same time.

AHTS are fitted with specialised equipment for better vessel manoeuvring. Modern AHTS have bow and stern thrusters, large

Figure 1.23 Offshore support vessel

deck space for storage and handling of anchors, anchor cables, buoys and associated gear including heavy winches. These vessels are considered to be the most powerful vessels in the category of tugs since they require significant 'bollard pull' to handle anchors. Furthermore, large storage areas for offshore structure supplies such as water, fuel and deck space for cargo make these vessels versatile and enable them to act as supply vessels as well. Additional uses include Emergency Response and Recovery Vessel (ERRV), installation of SPM moorings and buoys.

1.4.3 Platform Supply Vessels (PSV)

Modern PSVs are purpose built vessels whereas some older PSVs may be conversions from fishing or other smaller vessels. Generally, their length is up to 100 metres but smaller vessels can be found in near coastal areas.

As the name suggests, the main objective for these vessels is to transport personnel, equipment and cargo to and from offshore installations. The cargo carried include the consumables used on-board an installation e.g. drilling mud, fuel, potable and non-potable water, chemicals, food and provisions for workers on the installation. On their return passage, they bring ashore the materials that can only be disposed off safely in specialised reception facilities.

Figure 1.24 Platform supply vessel

Additional duties may mean that the vessel design has to be modified to achieve the required objectives. These modifications include the fitting of additional firefighting equipment and oil spill containment equipment. Some may be fitted with equipment capable of operating ROVs (Remotely Operated Vehicles) for underwater operations such as surveying, etc.

1.4.4 Emergency Response and Recovery Vessels (ERRV)

Developments in the offshore oil and gas sectors have led to the deployment of ERRVs. Primarily, their role is to 'standby' the off-shore structure so that recovery and rescue services can be provided in cases of emergency and subsequently survivors can be transferred to a place of safety.

PFEER (Prevention of Fire and Explosion, and Emergency Response) Regulations, 1995 regulation 17 requires the duty holder to ensure effective arrangements to:

- recover persons following their evacuation or escape from an installation;
- rescue persons near the installation;
- take such persons to a place of safety.

Figure 1.25 Emergency response and rescue vessel

In addition to the above, the ERRV can also act as an 'on scene coordinator' as per the emergency response plan of an offshore installation, act as a reserve radio station and warn off other approaching vessels to avoid risk of collision. PFEER Approved Code of Practice (ACoP), in addition, states that:

> There are many circumstances for which a suitable vessel standing by will provide effective arrangements and in these circumstances such a vessel will need to be provided.

Similar to PFEER, other guidelines and standards listed below also describe requirements for ERRVs. A few examples of such requirements are given:

- Oil & Gas UK's Guidelines for the Management of Emergency Response;
- OPITO approved Emergency Response Standards for ERRV crew training;
- Norwegian Maritime Directorate's Regulation No. 853 of October 1991 about Standby Vessels;
- OLF publication 016 – Norwegian Standard for ERRV operations. Note: OLF is now known as Norsk Olje & Gas;
- NOGEPA (Netherlands) industry guideline No. 6 – Code of Practice for Safety Standby Vessels.

The design of ERRVs also includes specialist medical facilities where medical care can be provided to persons recovered from water after an incident on-board the installation. Concurrent with design, these ships are required to carry personnel who have received specific medical training.

Whilst the OIM is responsible for safety of the offshore installation, the ERRV Master has the same role for their vessel. However, in cases where mutual cooperation is required, appropriate information must be exchanged and procedures established to ensure full assistance can be made available when required. In order to achieve this, the installation's duty holder must exchange information from the Emergency Response Plan (ERP) with the Master so that they can prepare for any emergency roles identified in the installation's ERP. In theory, the role of an ERRV can be compared with emergency service vehicles such as an ambulance or a fire engine which can be called upon in cases of emergency. However, the difference here is that the ERRV is required to be in the vicinity of the installation 24 hours a day and 365 days of the year. As a consequence of these requirements, ERRV crews must undertake additional training to deal with rescue operations (see Chapter 11).

Generally carrying a crew of 12 persons, the ERRV crew not only undertakes the normal STCW seafarer's training, but also obtains the certification below to meet the oil and gas industry requirements:

- Initial Shipboard Operations on ERRVs;
- Command & Control for Masters and Mates;
- Advanced Medical Aid;
- FRC Boatman's Course;
- FRC & Daughter Craft Coxswain's Courses;
- On-board Training & Development Programme for Masters and Crew.

(Details of the above courses are given in Chapter 11.)

Rescue and Emergency Response Equipment found on ERRVs include: fast rescue boat, daughter craft, dacon scoop, rescue basket, jason cradle, helicopter winch area or helipad, search light, emergency towing capability, fire fighting capability.

In addition to the above requirements, these vessels will be fitted with the navigation equipment and machinery as required by the classification society's requirements for the size and operation of the ship. However, some of these vessels are classed as multi-role vessels, in which case they can be capable of providing not only the ERRV operations but they can also be used as crew transfer vessels, supply vessels or even anchor handlers. In case of all these variations, equipment requirements will be dictated by the nature of operation and the area of trade.

1.5 Processing Oil and Gas

The hydrocarbons extracted from the seabed consist of a mixture of oil, water and gas. In many cases, they can travel from the reservoir through the wellbore to the surface under their own pressure. This process is known as 'natural drive'. However, in some cases, the reservoir pressure may be low at the start of extraction and/or reduce after some time. In this case, an 'artificial lift' has to be employed which may use in-well or seafloor pumps. If the temperature of extracted hydrocarbons is low, such as in polar regions, then oil or gas may need to re-circulate to force the hydrocarbons upwards.

When the hydrocarbon mixture extracted from wells reaches the surface, it has to be separated into three basic components i.e. oil, gas and water. Generally, these are passed through two stage separators to separate gas from liquids and then three stage separators to separate the three components.

Due to all components containing some water content, these require dehydration. The gas is then passed to a dehydration and compression system. Prior to exporting to storage or shore facilities, the gas may need to be treated with glycol to remove any residual water content. Finally, it is compressed and exported, whilst some of this gas may be used on-board the rig as fuel or for artificial lift.

The liquids go through a further separation of oil and water. Using the main oil line pumps, oil is exported to the onshore terminal or other storage facility such as an FPSO. The water is circulated into a water injection system after further treatment and mixed with seawater to maintain the reservoir pressure. Discharge of this water into the sea is normally not considered to be good practice as it may still have residues of hydrocarbons. However, if required due to operational reasons, it can only be pumped in waters of depths of around 50 m or more.

1.6 Utilities on Offshore Platforms and Rigs

In order to support platform operations, a number of subsystems or utilities need to be installed on the platform. These are described in the sections below.

1.6.1 Power Generators

Each offshore installation requires electricity to operate machinery and other utilities. Generally, diesel or gas turbine powered generators are used to generate this electricity. The number of generators will depend upon the power supply requirements but for rigs producing fuels, some fuel can be used by these generators. However, in many cases the latter option may not be economical. Even though natural gas is the preferred fuel for gas turbines, diesel or Liquefied Petroleum Gas (LPG) may be used as the main or alternative fuel.

During the initial installation of the rigs, diesel generators are used but subsequently gas turbines are installed for the continuous operation of generators. Gas turbines operate through the combustion of fuel or gas oil which in turn produces steam. This steam is then used to drive turbines connected with generator shafts that drive generators to produce electricity.

In order to meet the diesel or other fuel requirements, the rigs are fitted with fuel storage tanks. Offshore Supply Vessels (OSVs) are utilised to carry this fuel to the rig, which is then transferred from the OSV by means of hoses. The fuel system on-board a rig incorporates a treatment mechanism that utilises filters to remove water or other impurities from the fuel.

1.6.2 Flaring and Venting System

Flaring is a term used to describe controlled burning of excess or waste gas produced during oil and gas production operations. In

Figure 1.26 Flaring system on an offshore rig (Source: Rong Doi Project – Flickr)

some cases, it may be because the gas is just a by-product and in very small quantities that do not justify commercial production, hence it is burnt off using the gas flaring systems.

The flare system comprises a flare boom connected with pipes to collect the gas or vapours to be burnt. The system also utilises a specially designed 'flare boom tip' to improve the burning process by improving the mixture of air with gas being disposed off. Additional seals combined with a containing tank for liquids are installed in the system to prevent flashback.

In some cases, flaring may not be the required solution for excess gases such as carbon dioxide produced during the extraction of oil and gas. In such cases, the gas will be released into the atmosphere using a venting system. Sometimes, hydrocarbons mixed with other non-flammable gases may well be released into the atmosphere by mixing them with air to bring the concentration to safe levels avoiding the risk of ignition or explosion.

1.6.3 Seawater Systems

The seawater system can be used to supply water to a number of other components on a rig. The system consists of a number of pumps and filtering equipment to enable the required quality and quantity of water for various purposes. For example:

• Fresh water system – the main component of this system is the fresh water generator which is used to generate fresh water from seawater for on-board consumption. Fresh water generators may use either the reverse osmosis process to desalinate seawater or Vapour Vacuum Compression (VVC) distillation. Hypochlorite is used to control marine growth in the seawater system but when using a reverse osmosis generator, charcoal filters are used to remove traces of hypochlorites. The fresh water generator can be considered to be part of the potable water system which includes the purification and distribution of water to all areas of the rig e.g. water for showers and kitchens along with all pumps and accessories.

At times, fresh water may need to be imported via the OSVs or water barges and stored on-board in potable water tanks. Water brought in from outside or generated on-board will need to be chlorinated. Re-hardening filters may also be used to increase the pH value and make it potable water.

- Fixed firefighting system – this may be split into various components such as fire main and helideck foam system. Fire main provides water to various parts of the rig where fire hoses, deluge or foam systems may be in use.
- Cooling system – the seawater system provides a cooling medium to fresh water coolers used in hydraulic machinery, fresh water generator, generators and Heating, Ventilation and Air conditioning (HVAC). One of the main features to be considered for this system is the use of antifouling chemicals to prevent marine fouling of the system. The depth of cooling water intake and discharge are carefully selected to ensure the two waters at different temperatures are not mixed and the use of chemicals is reduced to a minimum.
- Washing facilities – these facilities require treatment of grey and black water from showers, toilets and kitchens before being discharged to sea.
- Water injection to maintain the reservoir pressure.
- Ballast system – this keeps the rig at the required draught to ensure adequate stability. The ballast system consists of specially allocated ballast tanks which are pumped in/out by using ballast pumps.
- Bilge system – this system is used to remove waste water, mainly a mixture of oil and water from the machinery spaces, pump rooms and other compartments. Bilge pumps and lines are kept separate from other pumping systems to ensure continuous availability and avoidance of mixing with other materials. The bilge system is required to be connected to the largest seawater pump to enable the fastest pumping in cases of emergencies.

1.6.4 Sand Separation System

The sand produced from the reservoir during the extraction process has to be separated from the hydrocarbon fluids and gases. This sand may be contaminated with hydrocarbons. The hydrocarbon content in this sand will depend upon the characteristics of the reservoir but in any case, the removed sand is required to be transported ashore for disposal. Alternatively, some facilities may re-inject this sand into a disposal well. Disposal into sea is not considered an industry good practice due to possibilities of causing pollution.

1.6.5 Drainage System

Drainage water is divided into two categories, open drains and closed drains. Open drains contain water from non-process areas whereas closed drains collect water from process areas where the water may be contaminated. The drainage system on the offshore rigs consists of drains for hazardous and non-hazardous materials. The system is designed to ensure minimal quantities of hydrocarbons are discharged during operations. In the UK, the Offshore Petroleum Activities (Oil Pollution Prevention and Control) Regulations define how these discharges are controlled. An example of such control is that the drainage water from decks including that from precipitation or seawater spray or equipment cleaning is required to be treated prior to being disposed of into sea.

1.6.6 Sewage System

Sewage from offshore rigs can be divided into two types – grey and black water with grey water being the drainage from showers and wash basins in the kitchen and laundry. Black water on the other hand is drainage from toilets or discharges from the hospital area which is passed through sewage treatment units before being allowed to be discharged at sea.

The MARPOL 1973/1978 convention definition of a ship includes 'floating craft and fixed or floating platforms'. Therefore the requirements under MARPOL for sewage treatment or for any other areas relevant to prevention of pollution apply equally to offshore installations. For sewage the basic equipment as well as the operational measures set out for ships also apply to offshore installations. MARPOL Annex IV gives the 'Regulations for the Prevention of Pollution by Sewage from Ships'. Its main requirements applicable to offshore installations are:

- Discharge of black water is permitted at a distance of more than 12 nautical miles from the nearest land.
- Comminuted or disinfected black water can be discharged at a distance of more than three nautical miles from nearest land.
- If an approved sewage treatment plant is used, then sewage can be discharged anywhere at sea.
- If the sewage contains mixed wastes, then the discharge should be according to the contents of the mixture where relevant requirements of MARPOL will apply.

1.6.7 Corrosion Protection System

The most common method for corrosion protection for offshore steel structures utilises cathodic protection by using sacrificial anodes for both the underwater and above water parts of the structures. A sacrificial anode may be manufactured from zinc and/or aluminium and attached to the steel structures. When connected to a steel structure, the cathode corrodes instead of the protected metal. The same results can also be obtained by using DC current through metal such as long pipelines. Such systems are known as Impressed Current Cathodic Protection Systems.

1.6.8 Helifuel System

Many rigs use helicopters as the main means of transportation to and from the installation's helideck located on the uppermost deck of the rig. As a result they may be required to have storage facilities for helicopter fuel. These facilities will consist of fuel storage and dispensing facilities. Components of this system include fuel filters, flow meter and hose reels with dispensing nozzles. Due to the nature of this operation, the system incorporates special equipment for electric bonding and a spill containment foam based fire fighting system. The helideck foam system consists of automatic oscillation foam monitors and a deluge system for the helifuel dispenser. When required, the system operates simultaneously to ensure all areas potentially affected by fire are covered. Personnel using this system are required to undergo specialised training.

Figure 1.27 Refuelling helicopter offshore

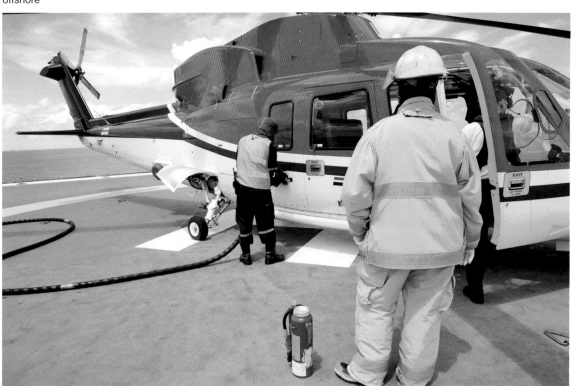

1.6.9 Inert Gas System (IGS)

An inert gas, also known as a noble gas, is a gas that does not react with other materials to cause combustion under certain conditions. Nitrogen, a colourless, odourless and non-hydrocarbon gas, is used as an inert gas in the offshore industry and is generated by specialised generators on-board using dry compressed air. It is mainly used during the drilling and completion phases of oil and gas wells. Additional uses include instrument panel inerting, flare gas inerting and pressure systems purging. Nitrogen can also be used to inert tanks where hydrocarbons are stored e.g. in FPSOs. This process is called blanketing. On FPSOs and oil tankers, inert gas is generated by using an IGS consisting of an inert gas generator that utilises fuel oils.

1.6.10 Accommodation

Following the Piper Alpha disaster, an offshore installation's accommodation module is kept as far as possible from the drilling and production facilities to enhance the safety of workers. In the UK, HSE Operations Notice 82 of April 2010 provides guidance for the provision of accommodation offshore as required by regulation 12 of the Offshore Installations and Wells (Design and Construction) (DCR) Regulations 1996. A summary of these requirements is given below:

1. Each installation must have a sufficient number of beds or bunks for personnel expected to sleep on it. This requirement also applies to Normally Unattended Installations (NUIs) if sleeping on-board is required at any time.
2. Hot bunking, i.e. two or more persons using the same bed/bunk to sleep in shifts, is not allowed.
3. A temporary increase in the number of persons on-board such as during refits offshore, prolonged maintenance shutdowns or other major work needs to be planned in advance to ensure sufficient number of beds are available at all times. Temporary beds are not allowed as they do not provide the required comfort and privacy to the occupants.
4. The regulations define an ideal room as one where at least 11 cubic metres space is available per person which includes bed, desk and wardrobes. Toilets/washing facilities are not included in this space allocation. The height of the space should not be less than 2.3 m. Any ceiling height of more than 3 m must not be included in volume calculation for personnel space. Any area providing less than 6.9 cubic metres space per person is considered to be overcrowded in which case enforcement decisions can be made by the inspectors who may require the number of personnel to be reduced to clear overcrowding.

5. In order to meet the above requirements, single or double occupancy cabins are considered satisfactory.

Notes

1 OGP (2014) International Association of Oil & Gas Producers – Fact Sheet. Available at www.ogp.org.uk/fact-sheets/ [Accessed 10.05.14].

2 A Brief History of Offshore Oil Drilling, Staff Working Paper No. 1. BP Deepwater Horizon Oil Spill Commission. Available at www.cs.ucdavis.edu/~rogaway/classes/188/materials/bp.pdf [Accessed 08.05.14].

3 Tyler, P. (2007) The Offshore Imperative: Shell Oil's Search for Petroleum in Postwar America. Texas A&M Press.

4 Gunter, M.M. (2013) Fatal Injuries in Offshore Oil and Gas Operations – United States, 2003–2010. Office of Safety, Health, and Working Conditions, Bureau of Labor Statistics, US Dept of Labor. Available at www.cdc.gov/mmwr/preview/mmwrhtml/mm6216a2.htm [Accessed 10.05.14].

5 Oil & Gas UK (2011) UK Offshore Commercial Air Transport Helicopter Safety Record (1981–2010); Helihub (2014) Accidents. Available at http://helihub.com/tag/accidents/ [Accessed 11.05.14].

6 Freudenrich, Craig, and Jonathan Strickland. 'How Oil Drilling Works' 12 April 2001. HowStuffWorks.com. Available at http://science.howstuffworks.com/environmental/energy/oil-drilling.htm [Accessed 11.05.14].

7 OGP (2012) Decommissioning of offshore concrete gravity based structures (CGBS) in the OSPAR maritime area/other global regions. International Association of Oil & Gas Producers' Report No. 484. Available at www.ogp.org.uk/pubs/484.pdf [Accessed 23.05.14].

8 Mathiesen, R. (1989) The Tender Assisted Drilling – NPD Experience, Offshore Europe, 5–8 September 1989, Aberdeen, United Kingdom.

9 IMCA (2007) Guidelines for the Design and Operation of Dynamically Positioned Vessels. IMCA M 103 Rev. 1. Available at www.imca-int.com [Accessed 26.05.14].

2

Controlling Offshore Hazards

2.1 Regulating the Offshore Industry

The offshore industry is one of the most hazard prone industries due to the very nature of the work and the atmosphere in which this work is carried out. Regardless of the basis of hazards, every worker has a right to a safe workplace. This can either come through offshore operator managed self-regulation or by allowing a governmental legislative body to provide detailed procedures. Both these options are likely to cause resentment amongst the workers due to one reason or another. However, the legislature and offshore operator must operate together where the former sets the goals and the latter establishes procedures to achieve them safely and successfully.

This chapter looks at the legislation applicable to the offshore industry and the functioning of legal systems under the English law. Examples from the English law have been used in order to ensure unambiguous understanding, keeping in view that the law in other countries may vary. However, the essence of health and safety legislation is generally similar in the entire oil and gas industry around the world and the principles are comparable for the basics.

In the UK, the Health and Safety Executive (HSE) regulates work related risks to health and safety in shore based and offshore industry. In order to set its objectives for the industry, HSE is supported by the legislation given in Table 2.1.

It is important for all workers to understand terminology used in the legislation to enable them to understand the consequences of non-compliance, both individually and collectively. For example, if a worker suffers an injury at the workplace, the possibilities are that:

* The injury was caused by lack of appropriate procedures, equipment, training or PPE. In such case, the duty holder or employer will be held responsible for the injury and may be required to face a court case either under the criminal or the civil law.
* The injury was caused by incompetency of an employee. In this case, the employer may still be held 'vicariously' liable for the injury as it is his responsibility to ensure competent employees are assigned tasks.

Table 2.1 – Legislation applicable to the offshore oil and gas industry

Year	Legislation
1972	Offshore Installations (Logbooks and Registration of Death) Regulations 1972 (SI 1972/1542) as amended by the Offshore Installations (Amendment) Regulations 1991 (SI 1991/679). These regulations provide the necessary framework for record keeping on-board installations.
1973	Offshore Installations (Inspectors and Casualties) Regulations 1973 (SI 1973/1842) provide requirements for inspections of installations for the purpose of checking compliance with legislation and to investigate any incidents or accidents. Offshore installation owner's duties to support this function is also detailed in these regulations.
1974	Health and Safety at Work Act (HASAWA) 1974 requires the employer to: • ensure health, safety and welfare at work of all employees; • provide safe and free of health risk equipment and work systems; • provide information, instruction, training and supervision related to health and safety. The employee is required to: • take reasonable care of their own and others' health and safety including consequences of acts and omissions; • cooperate with the employer to enable employer to comply with his statutory obligations.
1987	Offshore Installations (Safety Zones) Regulations 1987 (SI 1987/1331) define requirements for setting up safety zones around installations.
1989	Offshore Installations (Safety Representatives and Safety Committees) Regulations 1989 (SI 1989/971). These regulations define the functions of safety representative and requirements for safety committees.
1989	Offshore Installations and Pipeline Works (First Aid) Regulations 1989 (SI 1989/1671) set up requirements for provision of first aid and associated equipment for the workers employed on offshore installations and any other structures of pipelines connected to these installations.
1992	Offshore Safety Act 1992 extends the application of Part I of the Health and Safety at Work etc. Act 1974 and details powers given to the Secretary of State for preserving the security of supplies of petroleum and petroleum products; and for connected purposes.
1995	Borehole Sites and Operations Regulations 1995 (SI 1995/2038) set requirements including health and safety documentation for any sites where bores are being drilled for exploration.
1995	Offshore Installations (Prevention of Fire and Explosion, and Emergency Response) Regulations (PFEER) 1995 (SI 1995/743) provide requirements for protecting persons on the installation from fire and explosion; and ensure effective emergency response.
1996	Offshore Installations and Wells (Design and Construction) Regulations (DCR) 1996 (SI 1996/913) contain requirements for ensuring that offshore oil and gas installations, and oil and gas wells are designed, constructed and kept in a sound structural state, and other requirements affecting them, for purposes of health and safety.
1996	Pipelines Safety Regulations (SI 1996/825) contain requirements relating to the design, materials, construction and installation of pipelines and related health and safety requirements.
1997	Diving at Work Regulations 1997 (SI 1997/2776) state requirements for persons diving at work and impose duties on persons who control diving projects.
1998	Provision & Use of Work Equipment (PUWER) Regulations 1998 (SI 1998/2306): These regulations provide additional 'minimum' requirements for use of equipment including machinery, appliances, apparatus, tools or installation that may be used at work.

1999	Management of Health & Safety at Work Regulations 1999 (SI 1999/3242) cover risk assessment, principles of prevention, health and safety arrangements, health surveillance, hazard related information for employees. These regulations also cover employees' duties regarding safe use of equipment supplied by the employer and therefore place health and safety responsibility on employees as well as the employer.
2002	The Offshore Installations (Emergency Pollution Control) Regulations 2002 (SI 2002/1861). Under these regulations, the government is given powers to intervene in an incident where there is a risk of significant pollution or the operator has either failed or is likely to fail in implementing effective measures to prevent pollution.
2005	Offshore Installations (Safety Case) Regulations 2005 (SI 2005/3117) apply to both the mobile and fixed installations. A safety case is a set of arrangements designed specifically for one installation to minimise the risks and effects of a major accident. Risk assessments are required to show that all identified risks have been reduced to As Low as Reasonably Practicable (ALARP) level. The safety case must be approved by the HSE before the installation is allowed to operate. When approved, the safety case must be accessible to all workers on-board the installation.
2013	Health and Safety at Work etc. Act 1974 (Application outside Great Britain) Order 2013 (SI 2013/240) clarify requirements for activities undertaken in territorial sea adjacent to Great Britain and other areas designated under the Continental Shelf.

- The injury was caused by negligence of the employee. In this case, the employee may be held fully responsible for the losses and may need to share the consequences.

A full understanding of the requirements will enable workers to understand fully the consequences of their actions as well as their rights at any workplace.

2.2 Legislation

Legislation can be defined as the written, agreed and approved law known with common names such as statutes, acts or regulations. It may include new bills, amendments to the legislation through regulations, new acts related to law making and decisions of courts in the interpretation of existing law. Any law in its entirety cannot be developed in a short time; it takes years and goes through changes at all times due to social, economic, political and technological changes. The most common forms of the law are as follows.

2.2.1 Case Law and Equity

The law made by judges or juries when they interpret legislation to take decisions for certain cases for which there is no precise description or guidance in the existing law, case law is developed in two different ways and therefore has been given two names i.e. common law and equity.

The origin of common law in England is from around the eleventh century. Before the existence of common law, each small town, city or state used to have their local law meaning that even within one country the law was different for almost every place. In order to change this situation, the English kings appointed royal judges to bring about uniformity in the law throughout the country. Initially these judges had to travel to administer the law so as to keep it common everywhere. Since the idea was to apply a set of 'common' rules, the law thus implemented was given the name of common law. Even though its origins are from England other countries have also adopted its principles.

Equity is the type of common law which is utilised when the two parties to the dispute have not breached any law but there is a conflict. Historically, due to inequalities in the decisions made according to common law, the responsibility was delegated to the courts. Judges of these courts made decisions on the basis of even-handedness and fairness (equity) leading to established precedents. So in effect, equity supplemented common law and was applied in all civil courts, particularly in cases where there was a recognised conflict between common law and equity; the latter takes precedence. This process still continues with many other countries following the English system of equity.

2.2.2 Statutory Law

Statutory law is a set of rules in black and white (as opposed to common law) drawn up by a legislative authority or government to elucidate the working and running of government, maintain law and order or fulfil public requirements. In short, laws are made to form a set of rules which have a uniform application and which people can follow and get on with their daily lives. The statutory law is developed by members of parliament and approved by the head of state. Governments can amend the existing statutory law by introducing a 'bill' or they can develop new laws. The member(s) of parliament present a new bill, which is discussed in parliament, once passed it is drafted into law.

There are three levels of government in most democratic countries i.e. federal, state and local – all of them have some responsibility for law making. The basic responsibility for amending the law lies with the country's parliament but in some cases it is delegated to ministers and government departments. For instance, an Act if required may entrust authority for making detailed rules to bodies like the HSE, local councils or port authorities, etc. This type of law is called secondary, delegated or subordinate legislation that includes orders, rules and regulations, etc.

If there is any difference between the statutory law and the common law, the former has supremacy over the latter.

2.2.3 Oral Law

It is important to keep another form of law called *oral law* in mind, especially when considering health and safety at work. This is defined as a code of conduct followed by a group of people, a specific culture, race or religion. In this case, generally there are no written rules but instead the traditional values along with behaviours state rules that are more or less strictly followed. From the perspective of safety on-board offshore installations and vessels that operate in multinational environments, understanding of oral law has huge impact and significance on people's behaviours.

2.2.4 Legal Systems

2.2.4.1 Civil Law

This is the type of law that determines and deals with rights and duties between individuals and organisations on matters other than those that fall within the jurisdiction of criminal law. It covers affairs such as attributes of a person (e.g. name, age, date of birth/death, etc.), the relationship between individuals (e.g. marriage, adoption), property (e.g. boundary disputes, trespass, sale, purchase, possession), legal system of administration of relationship contracts (e.g. wills, sales, leases, partnerships). For example, a claim against the duty holder from an offshore worker during an accident will be dealt with under civil law provided that there has been no criminal activity or intent by any parties involved.

The civil law is made in a similar way to the criminal law by a mixture of statutes and case law. However, the civil law procedures are different and the court system separate from criminal law. In civil cases, the end result is usually to award financial compensation for damages or enforcement of a contract instead of any custodial sentence as in criminal cases.

2.2.4.2 Tort Law

The word 'tort' appears to have originated from the Latin word 'Tortus' meaning twisted or bent. In law, 'tort' is a branch of law that deals with compensation for losses such as those for personal injury, loss of earning, medical expenses or damage to property. The basis of claim under tort law is usually a civil wrongdoing with the aim to act as a deterrent for subsequent occurrence of similar acts as well as to place the sufferer at the same financial state as he or she was before the loss. However, the compensation for some torts can result in a criminal conviction and lead to imprisonment. The tort therefore falls into three general categories:

- *Intentional torts* are those in which somebody deliberately causes a harm, injury or damage to a person or property. For these torts the defendant is generally considered to know their duties but failed to perform them.
- *Negligent torts* are those in which someone fails to do what they were supposed to do due to their carelessness or their unsafe actions leading to harm.
- *Strict liability torts* are those in which the claimant seeks compensation for damages caused due to an unreasonably dangerous or unsafe product.

2.2.4.3 Negligence

In civil law 'negligence' is described as a breach of the extra-contractual responsibility on individuals and organisations for providing reasonable and prudent care. The basic notion of extra-contractual responsibility is not to breach any obligations causing injury to others in contrast to the contractual liabilities under which the statutory obligations required by terms and conditions of contract are to be fulfilled. In any dispute between two parties, a party can be considered negligent if it failed to exercise the same care that would have been exercised by somebody else under the same state of affairs. Therefore the first objective is to prove that the person owed a duty of care followed by a proof that the duty has been breached. In general, the negligence under civil law consists of five criteria:

2.2.4.3.1 *Duty of Care*

The exact limitations of the duty of care cannot be defined by law as it exists in almost all businesses, trades and even on residents of a house for their duty of care towards their neighbours where acts or omissions of persons can cause harm to others. Acts and omissions can be divided into two types:

- *negligent act* – in which a person fails to take appropriate precautions;
- *passive inaction* – in which a person does not take any action at all.

In all disputes in which duty of care has to be proved, the following defences may be used by defendant:

- *Voluntary risk acceptance*: If the defendant proves that the claimant gave an explicit or implicit agreement for the risk of damage, the claim may be denied.
- *Duty not owed*: If the defendant proves that they did not owe a duty of care, the claim may be denied.
- *Duty not breached*: If the defendant proves that they explicitly declared exclusion of liabilities claimed by the claimant.

- *Breach did not lead to damage*: If the defendant proves that the claim for breach by the defendant is not the cause of damage, the claim can be denied.
- *Contributory negligence*: If the claimant contributed to the negligence that led to harm (that would not have been caused otherwise), the claim is reduced by an amount proportionate to the claimant's contribution. For example if the defendant proves the claimant's 20 per cent contribution in negligence in a claim of £100,000 worth of damages, the defendant will pay only 80 per cent of damages (£80,000) of the total amount claimed.

2.2.4.3.2 Breach of Duty

Breach of duty occurs when a person fails to perform an explicit act or fulfil an obligation. In order to establish the breach, the following factors are considered:

- With an increase in possibility of magnitude of harm, the precautions also need to be increased.
- The precautions must be practicable.
- The precautions must be taken 'so far as is reasonably practicable' meaning that the risks should be considered against the benefits and cost (financial expenditure, time consumed and effort required) of risk control or prevention.
- If the precautions taken are according to the general practice of the trade, it is strong evidence that the defendant has not been negligent in fulfilling their obligations.
- If the defendant takes actions that are necessary in order to prevent a more serious accident, then a breach of duty may not occur; this defence is called 'defence of necessity'. An example is to pump out oil into the sea to save life, which would otherwise be a breach of law.
- Generally, in order to establish the breach, the differences in the skills of a professional are compared with an unskilled person under the same circumstances to establish the breach.

2.2.4.3.3 Occurrence of Damage or Injury

In order for a claim to be valid, the claimant must have suffered from physical or financial loss or both. The possibility of damage, loss in future or 'near misses' cannot be claimed. However, a psychiatric injury due to occupational stress can bring claims in the same manner as the physical or financial damage.

2.2.4.3.4 The Cause of Damage

All cases under tort of negligence must be proved by the claimant on the balance of probabilities where they have to prove that the harm was caused because of the 'breach of duty' on part of the defendant. The defendant, however, can explain the cause of harm which may provide a defence by proving contrary to the claimant's justification.

2.2.4.3.5 Legal Cause or the Proximate Cause

In many cases, there is more than one cause leading to damage. In any such cases where there are two successive causes of harm, generally the first cause is deemed responsible for the harm. The proximate cause or legal cause is slightly different in that the courts establish the primary or dominant cause amongst successive causes that lead to the harm. The term 'proximate cause' is quite important from an insurance point of view as the establishment of proximate cause determines whether the risk has been insured against or not.

2.2.4.4 Vicarious Liability

Under the doctrine of tort of negligence in civil law, another factor to consider is the employer–employee relationship when an employee may be the intentional or unintentional cause of a loss. This relationship is considered under the principle of 'vicarious liability'. The word vicarious has been taken from the Latin word 'vicarius' which means a substitute acting in place of another person. For example if a duty holder is held liable for the actions of any worker, the liability on the duty holder will be termed as vicarious liability, but in most of the cases it can only apply if the worker's action causes loss in the course of employment.

So the employer is vicariously liable for acts and/or omissions of the employee when they are at work because the employee is considered to be negligent due to a fault in the system set up by the employer. The claimant may show that the worker was not properly trained, monitored or supervised or was not informed about the possible hazards or risks. The employer, however, can defend by proving that either the employee was not negligent or he/she was acting on his/her own instead of being on the employer's business.

2.2.4.5 Burden of Proof in Civil Law

Each case brought in front of a court has at least two parties, i.e.

* claimant – the party (or parties) that has suffered a loss, damage to property or personnel injury and seeks compensation;
* defendant – the party (or parties) against whom the claim is brought.

In civil law cases, the claimant files the case and has to prove their case on the 'balance of probabilities' unlike in the criminal court where the proof has to be 'beyond all reasonable doubts'. It means that the burden of proof lies with the claimant in all civil court cases. However, because the objective in civil court cases is the award of compensation, a lesser degree of proof is required than that for criminal cases.

2.2.4.6 Criminal Law

The criminal law deals with enforcing the laws enacted by the state to ensure individuals or organisations do not break the rules governing society. If any rule or law is broken, they are taken as offences against the state and therefore are punishable even though they may not necessarily infringe the rights of an individual. A simple example of such a law is the traffic light signals i.e. if a driver drives through a red light on a traffic signal, this is punishable under the criminal law even if no damage has been caused.

2.2.4.7 Burden of Proof in Criminal Law

The claimant in criminal cases is called the 'prosecution' and in nearly all cases burden of proof rests with the prosecution. The proof has to be 'beyond reasonable doubt' which means that if a defendant demonstrates an element of doubt in the cases, they can be acquitted.

2.2.4.8 Mode of Trials

Usually in 'either way[1] (trial summarily or on indictment)' trial, a mode of trial hearing decides whether a case will be held in a magistrate's or crown court. Other modes of trial (summary trial or indictment) take place according to the provision given in the law.

2.2.4.8.1 Summary Trial

- In cases where the claimant and the defendant do not agree ONLY upon the final outcomes (they may agree upon the facts), the trials are conducted as summary trial.
- Minor offences such as breach of Health and Safety at Work etc. Act 1974 are tried under Summary Procedure.
- The trial is carried out by a justice (or a magistrate) without a jury.
- The maximum fine available in the summary trial system known as the 'statutory maximum' is currently £5000 but may be revised by the ministers of state.
- Generally summary-only offences do not carry the option of imprisonment but wherever it has been made available, the maximum is usually three months.

- At times, depending upon the severity of offence such as damage to environment by pollution or disposal of waste, the court can impose exceptionally high maximum fines (up to £50,000 in the UK) on summary conviction. This type of fine is known as 'Exceptional Summary Maxima'.

2.2.4.8.2 Trial on Indictment

This is a trial carried out by a judge along with a jury in the crown court (or sheriff's court in Scotland) or in the high court. The proceedings in such cases are known as 'solemn proceedings'. The judge gives decisions on points of law and the jury decides upon facts. The offences dealt with under this category are serious common law crimes such as murder, rape, dangerous driving or recklessly and knowingly causing damage. The maximum fine for offences dealt with under trial on indictment is an unlimited fine. In the UK, the maximum imprisonment given by the crown/sheriff's court is usually three years, but based on the seriousness of the crime, if the case is decided by a higher court the sentence may be life imprisonment.

2.2.5 Contract

A contract is defined as 'an oral or written agreement between two or more parties that the law will enforce'. It means that when parties agree, they create rights and obligations between each other. If a party to the contract is in breach of any of these rights or obligations, the other party can approach the court to enforce fulfilment of the contract or to claim for losses.

Contract law is generally derived from case law in most of the cases with some cases dealt with on the basis of 'equity' discussed earlier. In the case of international contracts involving two countries, it is usual to nominate one country whose law applies.

2.2.5.1 Types of Contract

Contracts can be of two types: informal (or simple) or formal contract. A formal contract is the type of contract that must be written in specific form in order for it to be enforced by law. All other types of contracts are called 'informal' contracts. However, it does not make any difference for courts when enforcement is in question as both formal and informal contracts will be enforced. The law requires certain types of contracts to be concluded in the given format due to information that must be included within them. Therefore the formal contracts to be signed by the parties are subject to the contract law and must have at least one witness with stated occupation. Simple or informal contracts on the other hand can be concluded in either writing, verbally or even implied by conduct in some cases.

Regardless of whether a contract is in verbal form or written form, if the terms and conditions are not complied with, the other party can claim for damages. However, each contract has certain terms and conditions that can be changed by agreement between parties. These are of two types:

- *Expressed terms and conditions*: those that are clearly spoken, written in a contract e.g. amount of money to be paid for renovating a house, freight for carriage of goods, wages for an employee, etc.
- *Implied terms and conditions*: those that are not spoken or written in a contract because they do not need to be stated. For example for carrying out a certain job, it is implied that the house must be available for renovation; the goods must be supplied for carriage; and the employee must perform the allocated tasks to earn wages.

2.3 Major Incident Hazards

2.3.1 Hazardous Offshore Environment

According to Offshore Safety Case Regulations 2005 (SI 2005 No. 3117) a major accident means:

1. a fire, explosion or the release of a dangerous substance involving death or serious personal injury to persons on the installation or engaged in an activity on, or in connection with it;

Figure 2.1 Offshore hazardous environment (Source: Richard Child – Flickr)

2. any event involving major damage to the structure of the installation or attached plant or any loss in the stability of the installation;
3. the collision of a helicopter with the installation;
4. the failure of life support systems for diving operations in connection with the installation; or
5. any other event arising from a work activity involving death or
   serious personal injury to five or more persons on an installation
   or engaged in an activity connected with it.

2.3.2 Hazard Examples

Since hazards can exist in any process or place, it is important
that all processes and places are carefully studied to identify these
hazards to ensure required objectives are achieved safely and successfully. Further details of hazard identification will be discussed
with risk assessment procedures later in this chapter.

2.3.2.1 Hydrocarbon Release (HCR)

Unintended HCRs account for almost half of all incidents in the
industry. These usually have a potential to lead to major accidents,
especially if a source of ignition finds itself in close proximity to the
released hydrocarbon. HCR can occur due to any of the following
reasons:

1. pipeline failure;
2. spillage from storage tanks;
3. loading of oil tankers;
4. loss of well control, or failure of a well barrier;
5. failure of a safety critical element;
6. significant loss of structural integrity, or loss of protection
   against the effects of fire or explosion;
7. vessels on collision course and actual vessel collisions with an
   offshore installation.

In most of the installations, sub-sea isolation valves are installed as
a safety barrier to shut down a flow line in case of an HCR.

The HSE's Offshore Safety Division (OSD) maintains a
Hydrocarbon Releases Database System where any HCR incidents are reported under the Reporting of Injuries, Diseases and
Dangerous Occurrences Regulations 1995 (RIDDOR). The HCR
severity is categorised as follows:

- Major HCR – a release that has a potential to impact on the
  outside of the local area such as on the escape routes and escalate to other areas of the installation, accompanied with serious
  injuries or fatalities. For example, a major leak which if ignited
  would lead to a major accident.

- Significant HCR – a release that has a potential to impact on the local area and cause damage to structure possibly accompanied with serious injuries or fatalities. A significant HRC can escalate to a major HRC if not controlled quickly.
- Minor HCR – a release that has a potential to cause serious injury to personnel in the immediate vicinity of the release but without any potential to lead to injuries or fatalities in surrounding areas. For example a minor leak that can be contained quickly.

The HCR incidents can be subdivided into the following categories on the basis of phase of operation in which they take place:

1. Hydrocarbon releases during drilling can occur in the form of:
   a. oil or gas flow from a well during drilling process known as drilling blowouts;
   b. accidental HCR when drilling or test equipment is being used commonly called as drilling spillage;
   c. hydrogen sulphide (H_2S) gas release at the drilling floor;
   d. mud spills during mud circulation in the well.

2. Hydrocarbon releases during production and export in the form of:
   a. accidental leakage or flow from a well during production referred to as 'production blowout';
   b. generally small in quantity but still with a potential to cause damage, spills from wellhead equipment called wellhead spills;
   c. release of hydrocarbons during separation, metering or pumping known as process leaks;
   d. fuel oil leakage such as that from diesel brought for on-board use.

3. Helicopter accidents: These generally occur during transportation of personnel to and from the offshore installations. Further details about helicopter travel and related issues are given in Chapter 9.

2.3.2.2 Utility Area Hazards

Utility areas such as the radio room, accommodation, galley, laundry or even the control room may become hazardous areas due to an accumulation of released hydrocarbons and consequently result in an incident if combined with a source of ignition. The galley is quite prone to pan/fat fires; living quarters can easily provide a source of ignition due to smoking. Machinery spaces contain all the necessary elements such as fuel and electric equipment to start a fire.

Figure 2.2 Confined space (Source: Petrofac Training)

2.3.2.3 Environmental Hazards

Environmental hazards such as movement of the seabed on which a well is located may cause failure of pipelines and/or associated equipment. Anchors or other mooring equipment securing the installation in place may be dislodged. Extreme weather may cause other problems such as structural damage.

2.3.2.4 Confined Spaces

Also referred to as enclosed spaces, any space that needs to be kept closed for operational reasons because of which it poses a risk of death or serious injury due to presence of harmful materials or unsafe conditions such as lack of oxygen within the space is classified as a 'confined space'. Some of the confined spaces can be readily identified as they will have limited openings. Examples include storage tanks for fuels, fresh or ballast water, etc. Some confined spaces may be less obvious but may contain equally harmful materials or conditions, for example store rooms that are not regularly used or ventilated, compartments adjacent to areas where explosive gases may be present. Historically, accidents have caused death whilst personnel attempted an entry into a confined space without adherence to approved entry procedures. The reason for these deaths has been 'complacency' on the part of the workers rather than the lack of procedures. In order to enter any 'confined space', an authorised competent person must check the atmosphere and certify for:

- sufficient quantity of oxygen in the air;
- absence of harmful gases, fumes or vapours.

In addition to the above, the space being entered must:

- be structurally sound to allow safe entry/exit;
- be free from trip/slip and fall hazards;
- be free from any leakages from adjacent compartments;
- provide suitable working conditions such as illumination, temperature and ventilation.

Further details about the hazards associated with entering any confined or enclosed space are given in the section on risk assessments.

2.3.2.5 Vibration Hazards

Workers subjected to long-term vibration may suffer from damage to their hands/fingers or other parts of the body. The most common problem in the industry is hand-arm vibration. HSE guide INDG 175 provides detailed information about the impact of hand-arm vibration. The two forms of permanent ill-health caused by this are:

- Hand-Arm Vibration Syndrome (HAVS);
- Carpal Tunnel Syndrome (CTS).

In order to ensure workers' safety, the duty holder is required to assess all risk, provide information and training to all workers who may be exposed to the risks and ensure health surveillance. In addition to the application of other legislation, the Control of Vibration at Work Regulations 2005 (SI 2005 No. 1093) have established requirements for exposure to vibration. Some important considerations with respect to these regulations are:

- daily exposure – the quantity of mechanical vibration to which a worker is exposed during a working day, normalised to an eight-hour reference period, which takes account of the magnitude and duration of the vibration;
- Exposure Action Value (EAV) – the level of daily exposure for any worker which, if reached or exceeded, requires specified action to be taken to reduce risk;
- Exposure Limit Value (ELV) – the level of daily exposure for any worker which must not be exceeded;
- hand-arm vibration – mechanical vibration which is transmitted into the hands and arms during a work activity;
- whole-body vibration – mechanical vibration during any work activity which is transmitted into the body, when seated

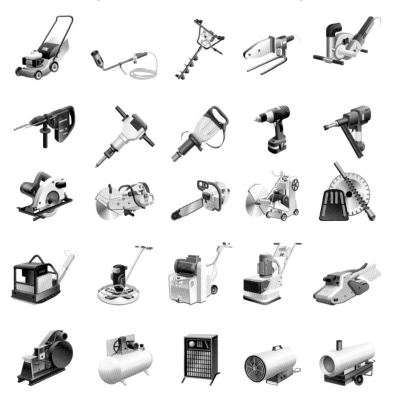

Figure 2.3 Examples of equipment causing vibration

or standing, through the supporting surface, during a work activity;

- vibration measurement unit m/s² A(8) – according to the British Standards, the unit used is the average acceleration expressed in m/s² for an eight-hour day expressed as A(8).

The prescribed EAV and ELV are as follows:

- For hand-arm vibration:
  - an EAV of 2.5 m/s² A(8) per day;
  - an ELV of 5 m/s² A(8) per day.
- For whole-body vibration:
  - the daily exposure limit value is 1.15 m/s² A(8);
  - the daily exposure action value is 0.5 m/s² A(8).

Duty holders are required to assess the risks considering:

1. the magnitude, type and duration of exposure, including any exposure to intermittent vibration or repeated shocks;
2. the effects of exposure to vibration on employees whose health is at particular risk from such exposure;
3. any effects of vibration on the workplace and work equipment, including the proper handling of controls, the reading of indicators, the stability of structures and the security of joints;
4. any information provided by the manufacturers of work equipment;
5. the availability of replacement equipment designed to reduce exposure to vibration;
6. any extension of exposure at the workplace to whole-body vibration beyond normal working hours, including exposure in rest facilities supervised by the employer;
7. specific working conditions such as low temperatures;
8. appropriate information obtained from health surveillance including, where possible, published information.

They are also required to provide information and training for:

- measures taken to mitigate any harm caused by vibration from tools or equipment used by workers;
- the exposure limit values and action values;
- the significant findings of the risk assessment;
- why and how to detect and report signs of injury;
- entitlement to appropriate health surveillance and its purposes;
- safe working practices to minimise exposure to vibration.

Manufacturers and suppliers are required to provide vibration emission information for the equipment supplied. This can be used by the duty holders or their nominated risk assessors to establish the exposure level. In addition, the supplier must also provide the following information:

- information about vibration related risks;
- training requirements;
- procedures for safe use of equipment including its limitations;
- information about vibration emissions.

There are two ways in which vibrations can be measured:

1. A specialised sensor can be attached to the equipment being used. This sensor is attached to a small computer that measures the real time vibration.
2. The vibration emission values supplied by the manufacturer can be used to estimate the impact of vibration on the basis of acceleration rates.

Workers are expected to follow the guidance below to avoid vibration related injuries:

- Seek familiarisation and training for the equipment supplied.
- Use appropriate gloves when using equipment that could cause vibration injuries such as hand tools. Gloves absorb some impact of vibration. In addition, never use hand tools with cold hands because of an increased risk of injury. Specialised anti-vibration gloves can be the most effective with high speed tools.
- Optimise work duration and shift pattern by seeking information about daily hand-arm or whole-body vibration exposure limits as calculated from the emission information supplied by the manufacturer.
- Regular breaks can give valuable time to body parts to recover from impact of vibration. It is recommended to take 5–10 minutes break after every 30 minutes' use of the tools/equipment that cause vibrations.
- Report any abnormal vibration in the equipment and avoid using it until the equipment has been checked.
- Ensure all the equipment goes through regular inspection and maintenance routine. Lack of maintenance can significantly increase vibration in most of the equipment.

Figure 2.4 Tools dropped from this height will injure someone (above). A simple solution can prevent this injury (below)

2.3.2.6 Dropped Objects

A large number of incidents related to objects dropped from a height continue to be reported in the offshore industry. It is estimated that dropped objects are one of the top ten causes of serious injuries and fatalities in the oil and gas industry. It is needless to reiterate that objects falling from a height will carry a significant risk of harm or injury to a worker underneath or damage to structures and equipment where the objects are dropped.

The dropped objects are divided into two categories:

1. Static dropped objects: objects that fall under their own weight from a static position. The reasons for this to happen include failure of fixtures or fittings such as screws, clips, nuts, bolts or gratings. It may also result from the fall of unsecured items such as tools, for example, a spanner falling from the hands of a worker. Some of these may result from environmental factors such as sudden jerky movement but other factors such as lack of inspection and maintenance cannot be ignored. This category of dropped objects is the most common in the offshore industry. In order to avoid this, the simple use of lanyards to secure tools/equipment or use of a tool belt should be considered.

2. Dynamic dropped objects: an object that falls from its position or separates from its main structure and 'flies' freely under the applied force, for example a tool falling from the hands of a worker (static dropped object), hitting a light on its way down causing it to break free and cause further damage. The light in this example will be a dynamic dropped object. Vibration in the equipment is also considered to lead to dynamic drop, hence must be carefully considered when assessing risks.

In essence, prevention of static drop will prevent dynamic drop as well. A careful risk assessment of the workplace and the tools being used will ensure this in addition to the use of appropriate means to secure tools and equipment.

2.3.2.7 DROPS Calculator

An organisation called 'DROPS' has been set up by a work group consisting of operators in the offshore oil and gas industry to raise awareness about implications of dropped objects and resulting damage. This organisation is running an ongoing campaign known as 'dropped objects prevention scheme'. Details of this campaign can be found on their website at www.dropsonline.org/. A database of incidents related to dropped objects, known as DORIS – Dropped Object Register of Incidents and Statistics has been set up as part of this campaign. DORIS has recorded over 850 incidents since it was set up in 2010.

DROPS offers software to benchmark potential consequences of a dropped object. A screenshot of the DROPS calculation carried out by this software is given in Figure 2.5 to highlight the difference in risk if a weight of 2 kg was dropped from a height of 10 and 1 metres.

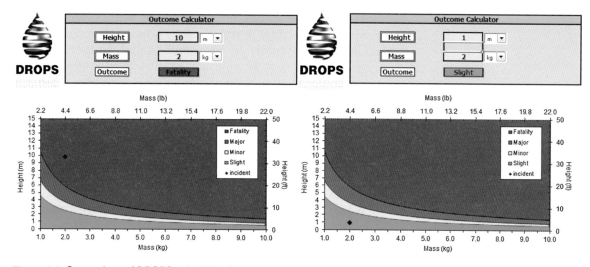

Figure 2.5 Comparison of DROPS calculation for a 2 kg item dropped from a height of 10 m (left) and 1 m (right) (Source: Screenshot from DROPS software downloaded from www.dropsonline.org/)

2.3.2.8 Dropped Objects Prevention

There are a number of steps that can be collectively taken by the duty holders and workers to prevent dropping of objects. These are summarised below:

1. Update the safety culture on-board to include information about measures to prevent dropped objects.
2. Inform all workers including contractors about the nature and scale of the problem and associated procedures in place.
3. Conduct hazard identification specific to the task paying particular attention to static and dynamic dropped objects.
4. Create safety zones and areas where it is unsafe to work whilst overhead work is being undertaken. DROPS guidelines provide three zones, green, yellow and red as described below:
   a. Green zone – an area where it is unlikely for workers to be exposed to dropped objects. Personnel may enter this zone if no additional restrictions are in place.
   b. Yellow zone – an area where workers MAY be exposed to dropped objects. Only authorised personnel should enter this area when required to perform a task.
   c. Red zone – an area with significant risk of workers' exposure to dropped objects. Only those personnel who are involved in a task linked to the dropping object can enter, i.e. these are the actual workers who are carrying out the specified task.

   Each of the above zones must have clearly defined boundaries as well as access/exit routes.

5. Agree and use specific signage for areas where overhead work with a potential for dropped objects is being undertaken.
6. Create an inventory of tools and equipment for any work being undertaken at a height so that it can be compared before and after work to ensure no tools have been left at the work site.
7. Ensure suitable Secondary Securing Devices (SSDs) such as lanyards and other accessories are available and used by all workers. This may require training individuals in their correct use.
8. Assess the use of secondary retention devices such as safety nets and implement as necessary.
9. When assigning tasks, allocate accountability to workers to ensure an ultimate ownership, particularly with reference to the dropped objects.

2.3.2.9 Noise

Noise is a known risk to workers' health which, if not prevented, may lead to permanent disability. The Health and Safety at Work etc. Act 1974 and the Control of Noise at Work Regulations 2005 (SI 2005 No. 1643) therefore require employers to mitigate any potential harm caused to workers by limiting their exposure to noise. Specific considerations to control noise levels on offshore installations include:

- minimise hearing damage to offshore workers;
- ensure normal communication can take place without any additional measures to enhance hearing;
- provide suitable ability to hear alarms and signals;
- ensure comfortable rest environment is available to workers in their time off.

Noise may be generated by power tools, machinery, explosives or impact of hammering, etc. Workers may suffer from temporary or permanent hearing loss. Whilst individuals can recover from temporary hearing loss in a short period of time, regular 'temporary' loss must be seen as a precursor to a permanent hearing loss which in almost all cases is gradual. Some of the symptoms of hearing loss include difficulty in distinguishing between the sounds of letters 't', 'd' and 's'. Eventually, this may lead to a condition known as 'tinnitus' in which people hear ringing, humming or whistling sounds without any obvious reasons.

Noise is measured in a unit called decibels (dB) often accompanied by letters A, or C etc. These letters indicate weighting to show average noise level. 'A' weighting is for average noise levels for human audible frequencies whereas B weighting is for audible frequencies' peak sound pressure exerted on human ears.

Figure 2.6 Noise and vibration can cause permanent damage

- A dB(A) of 0 is the lowest level of sound a person can hear.
- A dB(A) of 65 is for normal conversation level between different persons.
- A dB(A) of 85 or more can permanently damage hearing if exposed for a continuous 8 hours.
- A dB(A) of 100 can cause permanent damage in only 30 minutes of continuous hearing.

The Control of Noise at Work Regulations 2005 specify the level of noise to which workers can be exposed within any 24 hour or weekly period. It is important to understand the definitions below to comprehend various limits prescribed by these regulations:

- exposure limit value – a level of daily or weekly personal noise exposure or of peak sound pressure which must not be exceeded;
- lower exposure action value – the lower of the two levels of daily or weekly personal noise exposure or of peak sound pressure which, if reached or exceeded, require specified action to be taken to reduce risk;
- upper exposure action value – the higher of the two levels of daily or weekly personal noise exposure or of peak sound pressure which, if reached or exceeded, require specified action to be taken to reduce risk;
- weekly personal noise exposure – the level of weekly personal noise exposure, taking account of the level of noise and the duration of exposure and covering all noise.

The exposure limit values and action values stated in the regulations are as follows.

1. The lower exposure action values are:
   a. 80 dB(A) – a daily or weekly personal noise exposure;
   b. 135 dB(C) – a peak sound pressure.
2. The upper exposure action values are:
   a. 85 dB(A) – a daily or weekly personal noise exposure;
   b. 137 dB(C) – a peak sound pressure.
3. The exposure limit values are:
   a. 87 dB(A) – a daily or weekly personal noise exposure;
   b. 140 dB(C) – a peak sound pressure.
4. Where the exposure of an employee to noise varies markedly from day to day, an employer may use weekly personal noise exposure in place of daily personal noise exposure.
5. When considering compliance with the above values, due regard should be given to hearing protection supplied to and used by the workers.

Figure 2.7 An offshore worker with ear protection

Duty holders are required to:

- make hearing protection available to workers where they are required to work in high noise areas or upon their request;
- in the work areas where workers may be exposed to noise at or above the upper exposure action value:
  - declare the area to be designated a hearing protection zone;
  - identify the area by signs stating hearing protection must be used within the specified area e.g. machinery control room;
  - restrict the access to authorised personnel only;
  - carry out health surveillance on all workers exposed to noise;
- provide information and training for:
  - nature of risks from exposure to noise;
  - measures taken in order to mitigate the consequences of exposure to noise;
  - exposure limit values and upper and lower exposure action values;
  - significant findings of the risk assessment;
  - the availability and provision of personal hearing protectors and their correct use;
  - why and how to detect and report signs of hearing damage;
  - entitlement to health surveillance and its purpose;
  - safe working practices to minimise exposure to noise; and
- manufactures and suppliers to provide the end user with information about the level of noise created by the equipment supplied.

2.3.2.10 Slips, Trips and Falls (STFs)

HSE's report[2] indicates that in 2014, falling from height was the most common cause of fatalities and slips and trips were the most

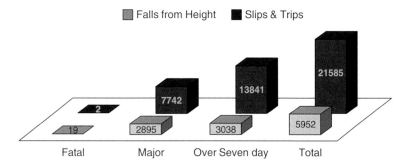

■ Falls from Height ■ Slips & Trips

Figure 2.8 STF incidents
(Source: HSE)

common cause of major injuries to the workers. Figure 2.8 shows the number of reported cases for STFs indicating that significant room for improvement still remains.

Generally STFs cause injuries to head, knees, ankles, feet, wrist, elbow, back, shoulder or hip. This is why most tasks either require physical measures to prevent a fall from any height or supply appropriate PPE to prevent injury. In order to understand fully the consequence of neglect to pay attention to STFs, it is important to know the differences in the meaning of three terms:

- Slip – losing balance due to loss of the foot's grip on the surface (deck/floor) due to lack of sufficient friction between the two surfaces, potentially resulting in a fall.
- Trip – catching an object with the foot or leg whilst walking, where the upper body continues to move forward and the lower body remains stuck, causing an unbalance, potentially resulting in a fall.
- Fall – movement of body from higher to lower level. Whilst with respect to the height of working surfaces, fall may be considered to be the movement from higher to lower level, but even at the same level, when the upper body moves towards ground, it comes under the definition of 'fall'.

Slips and trips can happen anywhere at the workplace, some of which can result in falls and give rise to even more serious consequences. These consequences can be divided into two general categories:

1. cost to the employer i.e. loss of productivity, costs of replacement employee and additional costs to rectify the damage and to care for the injured employee;
2. costs to the employee i.e. pain and suffering, lost income, medical or other expenses, possible disability and consequential impact on life style.

Types of injuries in the order of severity resulting from STFs are:

1. sprains and strains;
2. bruises or contusions;

Figure 2.9 Slips, trips and falls are the most common reason for offshore injuries

3. abrasions or lacerations;
4. fractures or dislocations;
5. death.

The Health and Safety at Work etc. Act 1974, Offshore Installations and Pipeline Works (Management and Administration) Regulations 1995 and Offshore Installations and Wells (Design and Construction, etc.) Regulations 1996 and other applicable legislation points to the same requirements, i.e. employers and employees must ensure health and safety … but the main question remains: how can this be achieved with respect to STFs?

Employers must carry out sufficient risk assessment to ensure the safety of workers. For example, the Offshore Installations and Wells (Design and Construction, etc.) Regulations 1996 state that:

- The floors of workplaces must have no dangerous bumps, holes or slopes and must be fixed, stable and not made of material which is or is liable to become slippery.
- When workers are employed at outdoor workstations, such workstations must be arranged so that the workers:
  a. are protected against inclement weather conditions and against falling objects;
  b. are not exposed to harmful noise levels;
  c. are able to leave their workstations swiftly in the event of danger or are able to be rapidly assisted;
  d. cannot slip or fall.

Not only to comply with legislator requirements but also to protect their own and workers' interests, the employer must therefore work with employees to assess the risks and take practical measures to prevent injuries. A few recommendations are:

Figure 2.10 Falls can lead to serious injury or even death

1. Clear walking surfaces that are free from contamination caused by wetness or stains from leakages around the area. Entrances for example may become slippery with personnel coming in with wet or greasy shoes. Simple door mats in addition to ensuring unblocked drains and scuppers could help overcome this problem.
2. Use of correct cleaning materials can be vital to remove any contaminants from the walking surfaces. Use warning signage to warn people of wet surfaces.
3. Spills should immediately be cleared with suitable cleaning materials.
4. Suitability of walking surfaces for maintenance must be considered. Any loose floorboards or gratings must be immediately fixed.
5. Sufficient illumination in all areas where personnel are likely to walk must be ensured at all times, especially in areas with steep bends or slopes.

6. For machinery spaces or other surfaces likely to be contaminated with oil, grease or other lubricants, suitable approved slip resistant safety footwear must be used.

7. Signpost STF hazards such as overhead cables, piping or other structures by using chequered or retro-reflective marking tape or fluorescent paint. Other simple warnings such as 'mind the step', 'trip hazard', 'slippery surface' etc. can prove very useful for avoiding STFs.

8. Safe walking practices such as holding on to rails where provided, particularly on stairs, can be very useful for protection against a fall. A recommended technique is known as 'Trailing Hand Technique' in which the strongest hand is trailed underneath the handrail behind the person when coming down the stairs. When going up, one hand should always be used to hold the rails. Using stairs with both hands occupied with something being carried is an invitation to a slip, trip or a fall.

9. Good task management is the key to safety in all aspects. Rushing through tasks not only reduces the time required to safely complete them but also diverts an individual's focus and hence leads to missing a safety related requirement. Remember! Safety comes at a cost, even if it is just to spend a little bit more time to concentrate on the task at hand.

Figure 2.11 Hazard warning signs

Figure 2.12 Fatigue can be overcome by sufficient rest

10. Good housekeeping ensures tools and equipment are safely stored away after completion of the shift or task. During work, all cables, ropes or other materials over which someone could trip must be tidied up and covered with trip prevention materials.

2.3.2.11 Fatigue

Fatigue can be defined as extreme exhaustion resulting from mental or physical labour or even illness. Some of the major oil spills such as Exxon Valdez in Alaska have been known to result from 'fatigue' in the workforce.

Offshore installations operate on a 24/7 basis where most of the operations need to continue without any breaks. If any operations are stopped, this may be because of breakdown of equipment/machinery or for planned maintenance. Workers on installations on the UK continental shelf are employed on 12 hours on/12 hours off shifts for a continuous 14-day period followed by a 14-day rest period. Different patterns of 14 shifts include 7 nights/7 days with a rollover shift in the middle or 14 nights/14 days for the whole duration. Consequently, the workers may not be able to get regular and restful sleep which can disturb their quality and duration of sleep and may eventually impact on their health and change their outputs. Fatigue induces lapses in judgement and therefore the ability to make required decisions. The oil and gas industry, however, enjoyed a very relaxed approach to this until recently, but the industry has realised this gap exists and is therefore constantly working to improve the situation.

Most people experience natural differences in their alertness during any 24-hour period. When workers are tired or in other words fatigued, they need to rest their body to regain alertness. Failing this, the alertness level reduces leading potentially to undesired consequences.

The two systems in the human body that control sleep are known as sleep-wake homeostasis and the circadian biological clock or rhythm. Under normal circumstances, sleep-wake homeostasis reminds the body to fall asleep when determined by the level of fatigue and helps us remain asleep until the system determines sufficiency of sleep achieved to regain the level of alertness required.

The circadian rhythm on the other hand represents a 24-hour cycle of physical, mental and behavioural changes in response to light, darkness or environment around a person. The circadian biological clock drives circadian rhythms. This clock is controlled by a group of nerve cells in the brain. A number of activities such as hormone production, cell regeneration, body temperature, etc. are linked to this clock.

Any impact on the circadian rhythm is therefore likely to cause disorders. Although self-sustaining, the circadian rhythm can be easily affected by external factors such as daylight, workload, stress, etc. An example of a disturbed circadian rhythm is jet lag which occurs when the circadian clock of a person who has travelled through various time zones quite quickly becomes desynchronised with the external time zones. Symptoms of jet lag include sleepiness, reduced alertness, depression, etc. The same symptoms can be seen in offshore workers who are fatigued.

Research[3] shows that if a person remains awake for the hours given below, then:

- 17 hours = 0.05 per cent blood alcohol – risk of accident increases by two times.
- 21 hours = 0.08 per cent blood alcohol (UK limit for driving is 0.08 grams of alcohol in 100 ml of blood) – risk of accident increases by four times.
- 24 hours = 0.10 per cent blood alcohol – risk of accident increases by ten times.

Due to the foregoing reasons, it is important to establish causes of fatigue so that they can be mitigated. These can be divided into the following categories:

1. Sleep and rest
   a. Lack and quality of sleep: collectively, this can be called 'sleep deprivation' which can lead to a reduced level of alertness and therefore increased potential for accidents.
   b. Rest periods and quality of rest: rest periods between shifts must be carefully monitored and managed. Especially in 12 hour shifts, no overtime should be undertaken.
2. Food (timing, frequency, content and quality)
   a. Eating too little: a very light breakfast will diminish very quickly with a rapid rise in blood sugar level after the breakfast and the crash after a short time period. A good breakfast will keep energy levels stable until the next meal.
   b. Food intolerance: the digestive system in many people is unable to digest all the food that is consumed. Consequently, despite having a stomach full of food, the person feels weak and constantly tired, a symptom of fatigue. If there are no other reasons, then this could be the cause and hence must be checked by a doctor.
3. Psychological impact
   a. Boredom: living away from family, almost locked up in a restricted environment can sometimes cause boredom. Excessive boredom can eventually lead to lack of mental rest and hence fatigue.
   b. Repetitive work can lead to boredom and exacerbate fatigue.

c. Stress can occur from overwork, worry, illness or injury. Whilst the human body can withstand normal stress, excessive stress can lead to a reduction in the level of alertness and attentiveness to the task at hand. If stressors are not resolved, then this can lead to depression.

d. Depression is a mental disorder causing people to suffer from:
   - low moods;
   - loss of interest;
   - loss of appetite;
   - lack of concentration;
   - disturbed sleep;
   - loss of energy.

   Depression can result from any of the foregoing reasons.

4. Physical impact
   a. Noise and vibration causing lack of sleep or rest.
   b. Chemical ingestion – at times things like inhalation of chemical fumes such as paints can result in nausea, dizziness and fatigue. Similarly, regular intake of caffeine alters the body's chemical make-up which can lead to fatigue if the body is not fed these at the required intervals.
   c. Jet lag can kick in when workers have to take long haul flights to reach their workplace.
   d. Excessive workload can lead to both mental and physical fatigue. Careful management of workload can help overcome any associated problems.

5. Medical conditions and medication
   Some medications such as antihistamines, cough and cold remedies can cause fatigue-like symptoms or aggravate fatigue. Their use should be considered in relation to the work environment and advice sought from the offshore medic or the safety officials on-board the installation. Some examples of medical conditions that can cause fatigue are:
   a. Anaemia: lack of red blood cells is known as anaemia. A person with this condition will almost always feel fatigued.
   b. Underactive thyroid: the thyroid hormone helps the body in converting food into energy. If it is not fully operational, the food remains undigested and therefore the body feels a lack of energy, another symptom of fatigue.
   c. Chronic fatigue syndrome: in some people, the symptoms of fatigue are persistent for periods of over six months. In such cases, a medical check-up is recommended.
   d. Diabetes is an illness where sugar remains in the blood stream instead of being converted into energy leading to a lack of energy and therefore fatigue.

   If the symptoms of fatigue don't disappear after taking operational measures, then a proper medical check-up should be undertaken.

In order to avoid the above complications resulting from fatigue, the HSE safety management model[4] based on the following is widely used in the UK sector:

1. Duty holder's policy for work shifts and schedules including objectives of the shift system, safety, alertness or performance and impact of shift work on circadian rhythm.
2. Established procedures and guidance for shift work and control of risks associated with fatigue. This guidance should cover all aspects of work and life on-board that could lead to fatigue.
3. Inclusion of human fatigue as a hazard when planning any tasks, carrying out risk assessments and implementing control measures. Examples of specific hazards that need to be considered are early shift start, overtime, off-duty call-outs, failure of being relieved from shift on time or at the end of trip due to extreme weather when helicopters can't operate and the resulting impact of hot bunking, etc.
4. Monitor impact of fatigue on the number of incidents/accidents, output at an overall level or at specific times of the day/night. This could be indicated also by complaints from workers, increased sickness/illness, use or excessive use of medication and tranquilisers, etc.
5. Inclusion of change requests, complaints or improvement notices in reviewing current procedures and practices.

It is worth noting that across merchant shipping and other industries, the hours of rest and work are considered in a completely different way in comparison with the offshore oil and gas industry. In the shipping industry, the hours of rest must comply with the following:

1. The hours of rest for each seafarer shall not be less than:
   a. 10 hours in any 24-hour period;
   b. 77 hours in any 7-day period.
2. Hours of rest may be divided into no more than three periods; one of which should be at least six hours long, neither of the other two periods should be less than one hour and the interval in between should not exceed 14 hours.
3. The legislation allows for certain exceptions such as due to call-outs, emergencies and drills for emergencies. In such cases, seafarers' rest hours must be compensated by additional rest hours for their rest time lost.

The above requirements, however, do not apply to the offshore oil and gas industry due to operational differences summarised below:

1. The Working Time Regulations 1998 (SI 1998 No. 1833) extended offshore through the Working Time (Amendment) Regulations 2003 (SI 2003 No. 1684) apply to the offshore workers.

2. Offshore workers mostly work on 14 days on/off rota allowing them to remain away from work for 26 weeks per year. Their work patterns may require them to work extra hours than the allowed hours.
3. The Working Time Regulations limit work hours to be averaged at 48 hours per week over a reference period of 52 weeks. These regulations allow the duty holder/worker to agree hours of work above 48 hours.

2.4 Health and Safety Guidelines for All Workers

Health and Safety (H&S) legislation and guidelines are all about protecting workers and ensuring their wellbeing. A key aspect of the foregoing is to highlight and thereby prevent the causes of ill-health or accidents through hazard identification and control. The primary legislation that oversees employee safety is the Health and Safety at Work etc. Act (HSW). It requires employers to ensure H&S of employees and employees to ensure their own and others' H&S. Generally, HSW requires a control over hazards by those whose work creates them, i.e. the employers or contractors.

Many offshore workers become ill or get injured in accidents at work in the offshore oil and gas industry. Whilst the pain and suffering of the individuals involved in these accidents is the prime concern, there are other costs such as impact on output, staff turnover, medical expenses and compensation claims. The requirements to avoid serious and imminent danger as required by the Management of Health and Safety at Work Regulations 1999 (Amended 2003), are given below.

1. Every employer shall:
   a. establish and where necessary develop appropriate procedures to be followed in the event of serious and imminent danger to persons at work in his undertaking;
   b. nominate a sufficient number of competent persons to implement those procedures in so far as they relate to the evacuation from premises of persons at work in his undertaking; and
   c. ensure that none of his employees has access to any area occupied by him to which it is necessary to restrict access on grounds of health and safety unless the employee concerned has received adequate health and safety instruction.
2. The procedures shall:
   a. so far as is practicable, require any persons at work who are exposed to serious and imminent danger to be informed of the nature of the hazard and of the steps taken or to be taken to protect them from it;
   b. enable the persons concerned (if necessary by taking appropriate steps in the absence of guidance or instruction and

in the light of their knowledge and the technical means at their disposal) to stop work and immediately proceed to a place of safety in the event of their being exposed to serious, imminent and unavoidable danger; and

c. prevent workers from resuming work where there is still a serious and imminent danger. This requirement may have some exceptions where remedial work is required to enable resumption of work.

3. A person shall be regarded as competent for the purposes of paragraph (1)(b) where he has sufficient training and experience or knowledge and other qualities to enable him properly to implement the evacuation procedures referred to in that sub-paragraph.

4. Spatial and situation awareness is important to ensure one's safety and that of others. Workers must be conscious of the operational equipment and work being undertaken around them. At times, simultaneous operations (SIMOPS) may be underway for various maintenance or operational activities for which workers must have knowledge through the use of PTW system.

Offshore work requires employee exposure to severe hazards given earlier in this chapter. Successful management of the offshore worker's H&S is based on five principles shown in Figure 2.14.

1. Employer's H&S Policy

Every employer who employs five or more workers is required by law to have a 'written' health and safety policy that defines standards and sets goals to be achieved. It also nominates specific personnel to carry out all required actions to achieve the aforementioned standards and goals.

H&S policy and personnel assigned various duties under it should not be separated from those with management functions. The policy must take into consideration all aspects of the 'work'

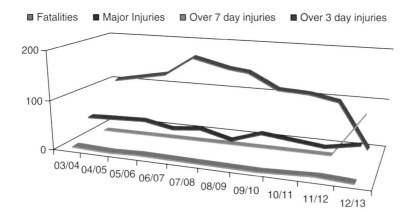

■ Fatalities   ■ Major Injuries   ■ Over 7 day injuries   ■ Over 3 day injuries

Figure 2.13 HSE accident statistics (Source: www.hse.gov.uk/offshore/statistics/hsr1213.pdf)

Figure 2.14 Management of H&S

to be carried out by the employees or workers including but not limited to, the following:

- worker recruitment;
- provision of work equipment, materials and workplace;
- procedural requirements to undertake any work including identification and control of hazards;
- top-down management involvement and commitment.

2. Safety Culture

Safety culture is defined as 'the product of the individual and group values, attitudes, competencies and patterns of behaviour that determine the commitment to, and the style and proficiency of, an organisation's health and safety programmes'.[5]

Human error is almost always a contributory factor for the majority of accidents. Safety culture is based on three factors i.e. human (competency and behaviour), machinery and system of work.

Safety cannot be achieved without considering the above three components together. For example, PPE should be used as a last resort when all other means to reduce/eliminate risk have been consumed. However, the use of appropriate PPE can only be ensured by ensuring that workers follow the guidance laid down in the procedures, with appropriate attention to the fact that the guidance or rules are comprehensive to include each possible aspect of health and safety.

Safety culture's goal is to minimise risks to an acceptable level. The industry employs personnel from different backgrounds including differences in language, religion and other social differences. Each human being is naturally 'biased' towards their own social values. For example, it may be thought that 'only European workers are highly professional' or 'third world country workers are inadequately trained'. These statements are obviously not true but a repeat of consistent behaviour leads to stereotyping that may have a negative impact on the safety culture. In order to overcome these issues, communication for all matters related to ensuring 'safety' must be effective. If the values, objectives, plans and views (even if contradictory) are made known through effective communication, then the differences can be overcome.

The key factor to achieve a safety culture is to recognise the fact that all accidents are preventable if correct procedures are established and followed by constantly thinking safety and continuous improvement. This can be done by following the three steps[6] below to implement a safety culture.

a. Top-down commitment

In any offshore workplace, the duty holder or the OIM has the ultimate responsibility for the safety of personnel, environment and the installation itself. However, no single person can deliver this responsibility without involving the whole workforce. But the involvement of all 'ranks' requires commitment from the employer, senior and middle managers as well as the workers. The requirements are summarised as top-down commitment and must consider factors such as:

- the owner or operator of an offshore installation must be involved through the duty holder in all aspects of operations;
- the role of on-board personnel must be recognised;
- those at the receiving end of instructions e.g. non-management personnel must be given opportunities for active involvement in all matters relating to safety.

b. Problem identification and safety monitoring

Constant monitoring of the work environment, particularly safety records, provides a means to identify problems so that they can be rectified. One of the main methods of monitoring used is commonly known as Lost Time Accident (LTA – when an employee loses work time for one day following the accident) based on which the Lost Time Accident Rate (LTAR – the accumulated number of LTA per million working hours) can be calculated. The LTA is used to report work related injuries or illnesses that have occurred during any given period of time. The LTAR of zero is for those work areas where there have been no accidents. The LTAR trend

Figure 2.15 Heinrich's triangle

can be used to see whether there has been any improvement in safety in the past and LTAR future outlook can be used for continuous improvement. This will ensure that there are no LTAs i.e. zero tolerance for accidents.

The adverse affect of an accident free place is that it may create a culture where workers become complacent and overlook changes that may lead to accidents. Heinrich's triangle[7] given in Figure 2.15 suggests that for every 330 unsafe acts, 30 will result in minor injuries and one in a major or lost time accident. It means that if problems are identified from near misses for example, the frequency of incidents that lead to them can be reduced and hence the chances of major injury can be eliminated altogether.

c. Affecting behaviour

Whilst the number of accidents and near misses is considerably reduced by continuous improvements in plant, place, competency and system, the personal attitudes of many employees still leave a small safety gap that needs to be fulfilled. Research suggests that most accidents[8] are an end result of a chain of actions that is instigated by an unsafe behaviour. It is therefore important to understand how behaviours affect the safety culture.

An example of behavioural effect in the workplace is use of PPE. Many workers use PPE merely due to peer pressure whilst the others may consider its use against their 'boldness'. Any such behaviours must be identified with a view to implementation of immediate corrective actions. The consequences of unsafe behaviour are generally not readily apparent; therefore it is difficult to affect behaviours. However, the means that can be used to improve behaviour are:

- Incentives – if an employee behaves as required, then he may be rewarded by a bonus, prize or promotion.
- Punishments – if an employee does not behave as required, then they may be punished by a fine, warning or may even be dismissed.

Additionally the safety culture must incorporate continuous safety thinking, monitoring of current practices and procedures, personal commitment of employees and employers. In short, affecting behaviour should be seen in conjunction with zero LTAR in the long term and is a continuous process that cannot afford any discontinuities.

3. Setting Goals

Careful planning plays a key part is setting 'correct' goals for any task, especially the tasks associated with ensuring the health and safety of workers. Plan essentially means the work required to achieve goals. The overall business goals must therefore be agreed by all stakeholders. Any agreed plans must be written down to ensure availability of appropriate records for legal purposes but also for anyone to revert to them to see the agreed scope of work. With respect to health and safety, these plans must include SMART[9] recommendations i.e.

- Specific to the issue;
- Measurable so that when the recommended action has been taken, the outputs can be measured against agreed goals;
- Agreed by all stakeholders including the personnel actually carrying out the job and the management, and comply with the legislative requirements;
- Reasonable keeping in view any limitations and the resources available;
- Time based – a time limit should be set for enforcement of any recommendations otherwise it may lose significance due to a change in circumstances.

4. Success Measurement

Success measurement is the study of reaching agreed goals through the agreed plans. Traditionally, the number of cases of injury or ill-health, LTA and any associated costs have been used to measure 'success' in any task or project due to the fact that all relevant records are readily available due to legal requirements. Such indicators to measure success in H&S matters are known as 'lagging indicators'. However, these may have some limitations because of which industry considers other methods to be more suitable. For example, if an action as a result of an accident to increase employee involvement in decision making was to increase their motivation, this can only come 'after' the accident and hence will be 'reactive' in contrast to 'preventative'.

Our objective should be to aim for an accident free workplace where the count for any LTA is zero. In such situations, lagging indicators will not exist in an ideal world. In order to ensure a proactive approach, a technique known as 'Positive Performance Measures (PPMs)' or 'leading indicators' is recommended. A leading indicator[10] is something that provides information to help users respond to changing circumstances and take actions to achieve desired outcomes or avoid unwanted outcomes. In doing so, it is desirable that actions will be taken to rectify any identified potential areas of weakness prior to an accident. PPMs can be divided into three types[11] shown in Figure 2.16:

- Activity (or input) measures where procedures supplied must be complied with. Any observed non-compliances must immediately be 'intervened' to prevent undesired outcomes.

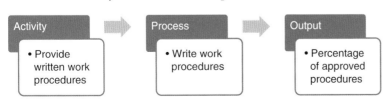

Figure 2.16 Measuring success

A record must be maintained to monitor behavioural trends in the workforce.

- Process measures to see the impact of activity measures. In recording non-compliances with activity measures, the non-compliant 'modified' procedures must also be recorded to observe a likely collective demand from the workforce requiring a procedural change.
- Output (action plan) measures provide an overall impact assessment on the 'end product'. This will define whether any changes are 'really' required and the possible resulting benefits and/or losses resulting from the change.

5. Training and Education

The offshore oil and gas industry is continuously evolving due to technological developments. Consequently, almost all personnel employed in this industry require training due to these changes as well as implementation of new regulations and for personal progress through promotions, etc. Some of this training can be delivered ashore but the on-board training can be divided into three main areas:

- Induction Training – this is required for all personnel joining an installation. However the level of training may vary with competencies as well the specific job role. Each individual will be required to complete certain tasks within short period after joining the installation.

  Generally, the following areas are covered in the induction training:

  i. company's health and safety, environment protection and pollution prevention policies and duties of individuals relevant to these policies;
  ii. organisation for safety on-board e.g. responsibilities for use and maintenance of Life Saving Appliances (LSA) and Fire Fighting Appliances (FFA) as well as safety officer/representative or medic's roles, etc;
  iii. procedures to deal with identified hazards including hazardous situations and/or equipment;
  iv. procedures to follow on hearing emergency alarms;
  v. requirements for emergency drills and relevant training on-board;
  vi. detailed procedures for dealing with emergencies giving details about alarms, muster stations, means of escape/evacuation (lifeboats, liferafts, Marine Evacuation System (MES), chutes, etc);
  vii. availability and use of PPE and other safety equipment;
  viii. location of first aid facilities, hospital and personnel responsible for them;
  ix. procedures for security of the installation.

- Emergency Response Training – ongoing training required for responding to on-board emergencies. Whilst all personnel are provided with 'generic' training as a pre-requisite to their employment, on-board training provides familiarity with specific on-board equipment and procedures.

- Health and Safety Training – the level of this training will vary according to the specified role i.e. the ordinary workers will have a different level of training as compared to a safety representative or a health and safety officer. In any case, the training for every employee must include familiarity with hazard identification procedures.

   The need for health and safety training can vary as a result of changes in legislation, equipment in use or an incident. As a general guideline, the following four steps need to be followed for any of the above areas of training.

Step 1: Educate – instruct and inform (orally or in writing) the person about the procedure, precautions and risks associated with a task. This is carried out as part of pre-employment training for all workers in the industry. However if they are delegated a task without any further training they may not perform as required due to lack of on-the-job training in compliance with local procedures.

Step 2: Train or coach – this is a coaching step where the information gained in the first step is put into practice. At this stage, guidance and support are both required for continuous improvement.

Step 3: Mentor – when adequate education and training is provided, the third step is to provide mentoring to

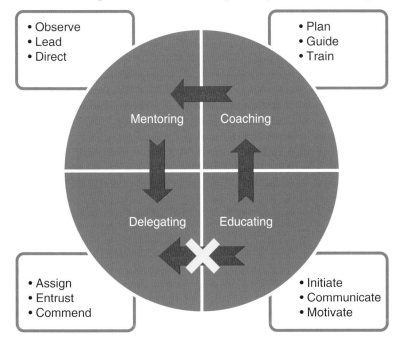

Figure 2.17 Safety culture in four steps

the trainee. At this stage, a nominated competent person should continue to provide guidance on the 'need' basis but should allow the trainee to perform the task. If required, further education or training can be provided here before proceeding to the next step i.e. delegation.

Step 4: Delegate – finally when the competent person is satisfied with the trainee's competence, tasks can be delegated without requiring any further intervention or guidance.

2.5 Safety Observation Systems

In discussing any system or its component that can help improve safety, human error appears to be always mentioned. Addressing the 'human' part of the work cycle to control the number of workplace accidents by addressing the human behaviour has therefore been the most important consideration for safety champions across the globe.

Human behaviour is based on the natural instincts of individuals where they may choose to find an easy escape from the problem or fight to overcome it. Instincts can be impacted upon only by cultural change through the practice of tested and established safe procedures. These may come through behaviour changes learnt through the environment around individuals, peer pressure or other social norms in the workplace. Generally, human behaviour is driven by financial motives, self-satisfaction or pride and dignity. These components combined with organisational attitudes such as existence (or non-existence of safety culture), management commitment along with perception and understating of risks as discussed previously in this chapter. A collective name given to this approach to create a system is known as 'Behaviour Based Safety (BSS)' systems.

Another aspect of human nature is 'attitude' which must not be ignored when BSS systems are being considered for implementation. Attitude can be defined as a 'person's internal state of readiness for action'. This state can be positively changed easily through reinforced and repeated exposure to a fresh motive. In short, when BSS theories need to be used to implement them as a system, both the attitude and behaviour changes need to be collectively analysed to see the successful results. Without this, the system will not produce the required outputs. As shown in Figure 2.18, conflicts can arise as a result of interaction between people. When seen in the context of safety, 'my attitude' can be translated into 'employer's attitude' which will change employees' behaviour. Now, this change can be positive if the employer (or others representing the employer, i.e. the management) is seen as acting in a positive manner, a salient feature of BSS. On the contrary, employees are quite likely to

Figure 2.18 Attitude and behaviour interaction

adopt a negative attitude which will obviously lead to a negative behaviour, a situation that contradicts the underlying concept for improvement of safety through BSS.

In order to improve safety, a systematic approach should be followed to address safety related issues after having identified them, but it is even more important to understand the basics. For example, when workers are employed by an organisation, the meaning of the term 'organisation' must be understood at all levels. W.J. Duncan defined an organisation[12] as 'a collection of interacting and interdependent individuals who work towards common goals and whose relationships are determined according to a certain structure'. It is evident from this definition that 'people' are the most important aspect of a workplace not the equipment or machinery. It is therefore important that the most critical part of an organisation performs safely to achieve the common good.

Behaviour Based Safety (BBS) systems, also known as BSS, Observation Based Safety Systems or Safety Observation Systems (SOS) were developed to improve individual workers' behavioural approach towards safety. In this system, workers are given responsibility to own their and others' safety in the workplace. Consequently, no unsafe acts or conditions are tolerated, and hence a vast majority of injuries are avoided by hazard elimination before hazards lead to accidents. Human error always appears to be a contributory factor in causing accidents but by effecting workers' attitude and approach, this inevitably results in a safer workplace.

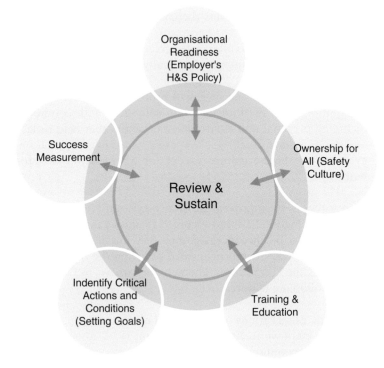

Figure 2.19 Behavioural based safety systems

The benefits that have been seen include increased 'worker' involvement in safety, improvement in individual and organisational performance.

The implementation of BBS is achieved simply through encouraging workers to get involved in raising safety awareness and intervening when unsafe acts or conditions are observed. Generally, if one worker considers an act or condition to be unsafe, it must be unsafe for some other workers based on the simple fact that if it is made safe for the most vulnerable of the workers, it will remain safe for the most competent persons. The key feature of this system is the 'avoidance' of blame culture in which every worker is supervising themselves and others. Instead, this approach helps in creating a 'safety culture' mentioned earlier in this chapter. Since the overall task of ensuring safety is split amongst the whole workforce, each individual employee shares a smaller workload whilst contributing it to the overall organisational objective of improving safety continuously. In addition, more effort is applied to accident prevention rather than to deal with corrective actions in a 'reactive' manner. After a successful implementation of BSS, it will maintain its effectiveness only when all system components continue to give peak performance, hence 'sustain' their outputs for a sustained and effective BSS.

Various organisations adapt BSS with a slight difference in approach. However, the common features for BSS are:

* understanding of the organisational structure;
* understanding of unsafe acts (behaviours/attitudes) and conditions;
* readiness to respond to all unsafe acts and conditions;
* just culture i.e. rules must apply equally to all without any distinction;
* continuous feedback and review of procedures and actions;
* recognition and reinforcement of safe acts and conditions including personal and environmental factors that lead to lack of risk comprehension;
* avoiding the blame culture except for appreciation of good work.

In essence BSS is an improvement of any existing H&S system. A comparison of Figure 2.14 and Figure 2.19 shows that all features of H&S have to be incorporated into BSS but through worker engagement and empowerment instead of giving the ownership to one person i.e. the H&S officer. The two-way arrows in Figure 2.19 signify the following features:

1. systematic observations of all acts and conditions;
2. employee empowerment and engagement;
3. personnel accountability at all levels;

4. support from management for all activities to develop sustainable mutual trust;
5. two-way recordable communication between employer and employee;
6. ongoing review and audit of all systems.

DuPont's STOP™ (**S**afety **T**raining **O**bservation **P**rogram) system is a well-known BSS in the offshore oil and gas industry. The salient feature of this system (and other BSSs) is the intervention into any unsafe acts or conditions. The system uses STOP™ cards for recording interventions and relevant details which can then be used to collage and analyse data. Other similar systems are:

1. TOFS – Time out for Safety;
2. Three 5s Observation System;
3. BP's SOC (Safety Observations and Conversations);
4. Chevron's POWER (Positive Observations Will Eliminate Risk) observation system;
5. ASA – Advanced Safety Auditing;
6. Care Plus;
7. Lock Out – Tag Out (LOTO) – It is a US Occupational Safety and Health Administration (OSHA) system based on The Control of Hazardous Energy (Lockout/Tagout), Title 29 Code of Federal Regulations (CFR) Part 1910.147. This system sets up the requirements to prevent inadvertent release of hazardous energy from equipment or machinery and therefore prevent harm from electrical, mechanical, pneumatic, chemical, hydraulic, thermal or other energy sources. Primarily, employers are required to establish a system through which employees can lock out all sources that have a potential to cause harm. If these sources cannot be locked out, then they must at least be tagged out to warn about the potential hazards.

Regardless of the system being used, understanding the importance of 'intervention' and keeping accurate records is the key to success of any system. In addition, a system in use must be capable of being deployed in any work area such as the drill floor, galley, production or storage and export units. As all workers are required to be involved in the system, the onus is on all workers and not only the supervisors or managers.

2.5.1 The Intervention

Interventions, for the purpose of BSS, are the actions that can produce the desired behaviour change. In other words, intervention is an interaction by ANY worker or stakeholder to ensure safety. This can be any programme or initiative that will lead to an improvement in safety.

safety observations and conversations

Date Occurred _____ Duration (in minutes) _____

Name(s) of Observers: _____

Type of Work Observed: _____

Key Safety Conclusions / Comments / Agreements Made: _____

_____

_____

Please enter Safety Observation Check only when applicable	Good Practice	Deviation	Comments
Plant			
1. Energy sources controlled	○	○	
2. Plant well maintained	○	○	
3. Leaks / spills contained	○	○	
4. Protection from hazards in place	○	○	
5. Housekeeping standard high	○	○	
6. Access / egress clear	○	○	
7. Layout and work locations safe	○	○	
People			
1. Work location protected	○	○	
2. Hazards understood	○	○	
3. Work positions safe	○	○	
4. People competent	○	○	
5. Distractions absent	○	○	
6. PPE appropriate	○	○	
7. Risk to others avoided	○	○	
Process			
1. Procedures valid	○	○	
2. Risks documented	○	○	
3. CoW Practice applied	○	○	
4. Correct procedures used	○	○	
5. Work well organised, systematic	○	○	
6. CoW requirements understood	○	○	
7. JRA / JSA participation	○	○	
8. Communication effective	○	○	
9. Change managed safely	○	○	
Performance			
1. Safety priority messaged by leadership	○	○	
2. Work pressure does not compromise safety	○	○	
3. Pace appropriate / safe	○	○	
4. Safety performance recognised	○	○	
5. Supervision appropriate	○	○	

Figure 2.20 BP SOC
(Source: BP plc)

SOC steps

Action
document in Traction, close gaps,
provide feedback

Conversation
reinforce good practice, risk assessment
style questions, coaching, agreement

Observation
triggers, energy sources, behaviours

Preparation
who, what, when, where?

Structured Conversation

Conversational risk assessment

The personal safety and process safety elements of SOC are, in effect, risk assessments in the form of a conversation. One key difference between the two elements is that the personal safety related solutions are generally either under the direct control or influence by the individual. In a process safety conversation, the solutions are more likely to be outside the control of the individual and an action may be taken forward by the leader.

task	hazard	risks	options	actions
- Task - Task environment	- Personal hazards - Task hazards - Plant hazards	- Personal consequences - Credible plant risk scenarios	- Preventative actions - Defences - Situation awareness - Error-provoking conditions	- Recognition - Commitments - Actions

Figure 2.20 (*continued*)

Interventions can be divided into two categories i.e. direct and indirect interventions. Direct interventions impact on human behaviour through force whereas indirect interventions impact upon an individual's thought and decision-making process. In other words, indirect intervention will change the attitude which will eventually lead to a change in behaviour. A compulsory change in any procedure is an example of a direct intervention, but on the other hand highlighting the consequence of a lack of adherence to tested procedures is an example of indirect intervention.

Interventions can be carried out at different levels of authority at any workplace. For example, interventions to improve policies, procedures or organisational structures may be initiated at management level whereas any changes to equipment may be recommended by workers at non-management positions. In any case, a Plan, Do, Check, Act (PDCA) approach is considered to ensure the maximum effect of the BSS. Various components of this technique are shown in Figure 2.21.

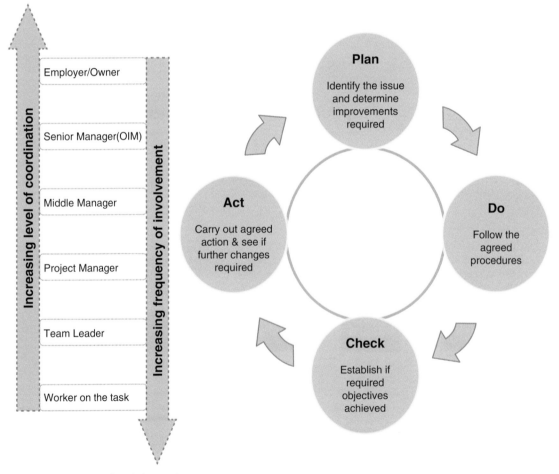

Figure 2.21 Plan-Do-Check-Act cycle

On a day-to-day basis, each worker must understand the general steps for an intervention based on observation of unsafe acts or conditions. These observations require individuals to make time in their normal routine to carry out interventions in a non-threatening environment. Observations must be both positive for good practices and negative for unsafe acts and conditions. The recommended steps for a successful intervention are given below for guidance.

1. Define: Keeping in mind that any person can carry out observation based safety intervention, the intervener must define the scope of intervention prior to taking any action. This may involve gathering all the necessary information to ensure all required data or information is readily available. The intervener must ensure that any work should not be stopped at a critical stage where further complications may arise as a result of stoppage.

2. Observe: The intervener can then stop the job, keeping in mind any specific requirements of the BSS in place. For example, in the '3 Fives' observation system,[13] the intervener may need to assess one of the following:
   a. Unsafe Conditions:
      i. eyes not on task
      ii. mind not on task
      iii. contact/striking/trapping
      iv. balance/traction/grip
      v. violation of permit/procedure/rule.
   b. Unsafe State of Mind:
      i. haste
      ii. frustration
      iii. fatigue
      iv. complacency
      v. wilfulness.
   c. Unsafe Outcomes:
      i. serious injury (multiple)
      ii. serious injury (one person)
      iii. first aid
      iv. material damage
      v. minimal.

3. Intervene: Once the intervener has stopped the task, they can then introduce themself and explain the purpose of intervention. The most important thing at this stage is to ensure avoidance of conflict and not to appear as if the intervener was trying to gain some personal benefits. Accusations must be avoided at all costs to ensure effectiveness of the system, remembering that the aim is to improve safety and not to penalise individuals for their mistakes. In doing so, the intervener must discuss (a two-way communication) the issue. It is not necessary that the intervener will always be 'right'. If they are found at fault, the

Figure 2.22 Safety intervention cycle

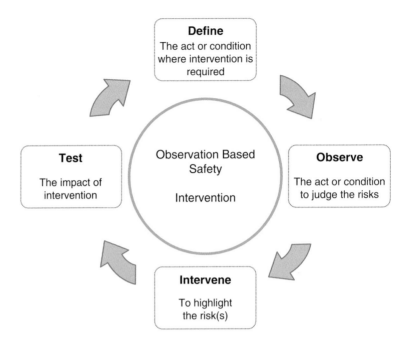

situation can be resolved easily as we all learn from mistakes. At the same time, if the intervention is for a good practice, then the person intervened must be praised. Remember! You may also be intervened by others. Some examples of the questions that may be helpful are:

a. Can you tell me exactly what are you doing?
b. Why do you think I have stopped you?
c. Do you know what could have gone wrong if I had not stopped you?
d. Can you elaborate the hazards you have identified?
e. Have the identified hazards been risk assessed?
f. Do you know the possible damage or injury that could have been caused?
g. Have you seen the Job Safety Analysis (JSA)?
h. Can you show me the written procedure for the job?

   Once the intervener has asked the required questions, the worker must be allowed to explain fully their side of the story, especially how they believed the job could have been completed without compromising safety.

4. Test: Any unsafe act or condition must be recorded on the observation card and reported immediately. Depending upon the nature of the task being stopped, if the fault or weakness can be rectified, then the work may resume; otherwise a further investigation may be required. Observations cards do not need to record the name of the person being observed, particularly if the intervention was due to something that could have otherwise led to an adverse outcome.

2.5.2 Personal Responsibility for Safety (PRfS)

Despite advances in technology, increased training and reduction in the number of accidents, a significant room for improvement in safety assurance still remains in the offshore oil and gas industry. Human error continues to be a contributory factor in most accidents. As a consequence, the industry has recognised that without impacting upon individuals' behaviours through giving them responsibility for their own and others' safety, it will not be possible to achieve further improvements. This concept introduced through the Step Change in Safety is known as PRfS and involves 'everyone' regardless of their position within any organisation. The PRfS approach is based on the following elements:[14]

1. **Clear Expectations**: Workers and employers' expectations vary with job roles. All personnel must be informed about expectations, particularly with reference to each individual's personal responsibility for safety (PRfS). Once they are informed about the basic requirements, they must also be informed about the consequences for lack of compliance. The salient feature of PRfS being the freedom to:
   a. intervene when any unsafe behaviour or practice is observed;
   b. share good practices to improve standards.
2. **Effective Communication**: Transmission, reception and understanding of the transmitted message are three key stages of effective communication. Transmission of any communication from managers does not necessarily mean safe and successful follow-up by the subordinates. At times, simple mistakes like use of jargon or abbreviations can lead to failure of communication. The human factor is a very effective consideration to ensure completeness of communication. For example, the body language or facial expressions and tone of communication may contribute to effective understanding of the transmitted message. Open conversations instead of authoritative one-way communication, particularly on matters related to safety, show involvement from all stakeholders in a workplace and hence lead to extremely useful contribution. One-way communication makes any job a chore with most employees losing interest in feeding back to management as they are not being seen as effective in problem solving.
3. **Personal Leadership**: An unignorable fact is the manager's duty to be responsible for the end result or product. It is here that the difference between a leader and a manager needs to be established. PRfS encourages personal leadership at all levels where each individual strives to accomplish tasks without any injuries. Management does their bit by ensuring provision of all resources to achieve the required results. Subsequent two-way

cooperation means openness and personal integrity of each person involved without the fear of being reprimanded.

4. **Personal Risk Awareness**: Encouragement for personal leadership nurtures an atmosphere of raised risk awareness where workers are not fearful of costs or pressure from senior management for raising their concerns about all matters related to safety.

5. **Planning**: Once the PRfS system is in place, appropriate personnel are identified to plan actions to rectify any identified problems. This is carried out by involving personnel who may be directly or indirectly affected by those problems. Consequently information flows in both directions i.e. top-down and bottom-up, promoting confidence at all levels in the workforce.

6. **The Right and Duty to Intervene**: Success of STOP™ or TOFS and other similar systems are significant examples of individuals being confident enough to intervene in any unsafe practice. Consequently, a safely culture has developed in workplaces which have promoted any such systems. Workers now consider it as their right to be stopped if they are not following correct procedures and their duty to intervene when they observe something similar.

7. **Accountability:** By definition, accountability is answering someone for your actions. Managers can delegate tasks and duties but not the accountability. Under PRfS, every individual is accountable for their actions or lack thereof. If any employee chooses not to perform their job safely, they can be held accountable for it. By involvement of all personnel in ensuring safety, everyone is responsible as well as accountable for their and others' safety. Effectively, this converts the whole organisation to a team and it is needless to highlight the effectiveness of teamwork for success.

8. **Self-evaluation:** PRfS is about openness which includes self-evaluation by presenting one's own thoughts and fears to others to obtain feedback and instil improvement in their practices.

9. **Develop, Encourage and Sustain Safe Behaviours:** By sharing weaknesses, individuals find opportunities to overcome their fears as well as learn from others. Help can only be offered if individuals express a desire to be helped. Once a culture of improving safety through mutual help and support is set up, it can only improve, provided support from management is maintained.

Further details about PRfS can be obtained from the Step Change in Safety website at www.stepchangeinsafety.net.

2.5.3 Incident Databases

Whilst every organisation has its own system for recording incidents/accidents, the industry also uses a number of databases to analyse information and study trends on a wider scale. Some examples are given below.

1. Incident Alerts Database – sponsored by the UK offshore oil and gas industry, it was previously known as SADIE (Safety Alert Database and Information Exchange). This database is designed to share information but the information available is not exhaustive enough to be used for statistical analysis. The information can be accessed through the Step Change in Safety website at www.stepchangeinsafety.net.

2. The UK's Health and Safety Executive (HSE) has a separate division known as Energy Division (ED) which is responsible for regulation of the offshore oil and gas industry alongside OSD (Offshore Safety Division). Since HSE is a government body, it maintains a comprehensive information resource on all aspects of offshore industry. Full statistical information can be obtained from www.hse.gov.uk/offshore/statistics.htm. In addition it also publishes a special newsletter 'Tea-Shack' for offshore workers. HSE and DECC (Department of Energy and Climate Change) in the UK have formed the Offshore Safety Directive Regulator (OSDR) which acts as a competent authority to implement requirements of the EU directives on safety of oil and gas operations in the offshore industry.

3. WOAD – World Offshore Accident Database is operated by the classification society Det Norske Veritas (DNV). Further information can be obtained from http://woad.dnv.com.

4. SINTEF Offshore Blowout Database – The acronym SINTEF has been taken from the Norwegian words 'Stiftelsen for Industriell og Teknisk Forskning' meaning 'The Foundation for Scientific and Industrial Research'. The SINTEF database contains information about worldwide offshore blowouts/well releases since 1955. Full details can be found on their website at www.sintef.com.

5. Well control incidents database – The OGP database used for recording well control incidents since 2012. The database is designed to capture both accidents and near misses. Further information is available at http://info.ogp.org.uk/Safety. Other information resource maintained by OGP is fatal incident and high potential event databases.

6. Energy Resources Conservation Board (ERCB), Canada maintains information about offshore blowouts in Canada.

7. Neal Adams Fire-fighters (NAF), Texas USA contains information about onshore and offshore blowouts in the USA.

2.5.4 Safe System of Work (SSOW)

The offshore oil and gas industry deals with large quantities of hazardous flammable and toxic substances which carry an extremely

Figure 2.23 A PTW being discussed
on site

high risk of accidents. To prevent such an occurrence, all processes need to be effectively controlled through safe systems of work.

A safe system of work (SSOW) is an agreed procedure designed after a careful systematic study of any task to identify all hazards and take measures to control the resulting risk prior to commencing the task. Various other names are used for SSOW such as Standard Operating Procedure (SOP), Local Operating Procedure (LOP), method statement or simply a Work Procedure (WP). Any SSOW will contain the key elements below:

1. Permit to Work (PTW);
2. Task Risk Assessment (TRA) or simply risk assessment;
3. task definition and scope;
4. toolbox talk;
5. tools and equipment required;
6. work schedule and rest hours;
7. awareness of any safety issues and procedures to raise alerts;
8. supervision.

2.6 Permit to Work (PTW)

A PTW is an integral part of any SSOW to record processes that manage any identified hazards and control associated risks in safety critical tasks. A PTW itself does not reduce incidents but it is a means to reduction of incidents. In addition, it is a formal method to communicate the required precautions and procedures to all relevant stakeholders, including management.

The industry uses slight variations in the name assigned to PTW such as a 'permit', 'work permit' or 'permit to work' but they all mean the same document that forms part of the SSOW. In the UK, a Task Risk Assessment (TRA) defines the tasks that will require a PTW considering:

- person(s) authorising particular jobs;
- identification of tasks or hazards where compliance with PTW system will be required;
- person(s) responsible for stipulating safety measures;
- requirements to train personnel using or involved in PTW system;
- responsibility for monitoring compliance with PTW system agreed procedures;
- record keeping;
- tasks being undertaken simultaneously;
- involvement from management, even if it is just to raise their awareness about the task being undertaken;
- times when any task needs to be stopped due to other conflicting tasks;
- a formal handover procedure between workers on different shifts.

PTWs are considered mandatory for the following tasks:

1. work at height including lifting persons with lifting equipment (see Chapter 4);
2. work on fragile surfaces such as roofs;
3. work in confined spaces;
4. work on evacuation, escape or rescue systems;
5. hot work such as cutting, welding, soldering or any other work where hot flames or grinders may be used outside of a designated workshop area;
6. work on fire safety systems such as alarms, fire or smoke detectors or their associated components (see Chapter 6);
7. work on electric supply systems or high voltage equipment;
8. work in areas where asbestos may be present;
9. work on equipment that may require breaking containment of flammable, toxic or other hazardous substance;
10. work with chemicals that fall under COSHH (see Chapter 5);
11. work in the vicinity of or with explosive, radioactive or highly flammable materials;
12. work on conveyors, lifts or mechanical hoists;
13. work on pressure systems such as pressure testing;
14. work on scaffold towers or mobile elevating work platforms (MEWPs) (see Chapter 4);
15. diving operations.

On each occasion when a PTW is to be used, it should be checked for its adequacy for the intended task considering the following factors:

- *Personnel* – personnel involved should be identified in the PTW. Means should be provided to check their competency for the task.
- *Communication* – all personnel involved including those required to monitor, review and provide assistance should be identified and informed.
- *PPE, Tools and Equipment* – any of these required for the task should be identified and provisions made to check adequacy.
- *Task* – information about the task itself should be contained in the permit. This information should include the nature of work, the duration within which it is intended to be completed and any contingency measures agreed.
- *Authorisation* – the PTW should be authorised by appropriate persons. In most cases, the OIM signs the permit to work. The sequence for obtaining signatures should be:
  a. the person(s) carrying out the task;
  b. the authorising person i.e. safety officer or other authorised officer;
  c. the OIM, area authority, area operator or team lead – one or all of these may need to be involved depending upon procedures agreed in SSOW.
  The above sequence ensures that each person involved in the operation has fulfilled their role as intended.
- *Review and Management* – the PTW should be continuously reviewed and managed by appropriate authorised and competent personnel.
- *Record Keeping* – a copy of the PTW should be kept for records as required under installation procedures.
- *PTW* – Each permit should be specific to the task.

A sample PTW[15] form is given in Table 2.2. This PTW should not be used without a checklist given in the next section on 'checklists'.

2.6.1 Issuing a PTW

The procedure for issuing a PTW may vary slightly between offshore operators. However, generally it is based on the following steps:

Step 1. Work Request: when the work request is received for a non-routine task, the originator (OIM, permit authoriser or area authority) considers the request including any other similar task being undertaken and begins the process for issuing the PTW.

Step 2. Task Scope: depending upon the requirements of the task, a careful consideration is given to understanding the requirements of the task, the personnel and their competencies required to carry out risk assessment and undertake the task. If the scope of a task changes, the PTW may need to be cancelled and a new one issued based on the new task scope.

Table 2.2 – Permit to work		
Permit to Work		Oil Rig 1 Ref: ....

Notes: The checklist relevant to the intended task should be filled in and attached to this permit. The sections not applicable should be crossed out or NA written within.

Task Details: ....
Task Location: ....
Plant Identification: ....
PPE/RPE Requirements: ....

Validity of Permit: From ...... *(Date & Time)*............. To........... *(Date & Time)*...........
Person Supervising the Task .... *(Name and Signature)*............................
Person(s) Performing the Task .... *(Name and Signature)*........................ ......
.... *(Name and Signature)*........................ ......
.... *(Name and Signature)*........................ ......
Reference Checklist: ....

Authorising Person	**OIM**
....	....
.... *(Date & Time)*..............	............*(Date & Time)*............

Completion Certificate

Note: This part should be completed and remarks entered as appropriate at either the completion of the task or at the expiry of the permit. In both cases, it should be signed by the authorising person, the task supervisor and kept in the file 'Permit to Work Records'.

Remarks ....
....

Supervisor	**Authorising Person**
....	....
.... *(Date & Time)*..............	.... *(Date & Time)*............

Extension/Shift Handover: *(state particular hazards or other information necessary to ensure safety during this phase)*
....
....

Cancellation/Suspension: *(state reasons, in case of cancellation due to completion of the task, complete 'Completion Certificate' above)*
....
....

Step 3. Classify Task: after carrying out hazard identification and understanding details of the task, a consideration is given to whether existing procedures are sufficient. If they are not, then new procedures need to be agreed. At this stage, it may be required to divide the task into smaller easily manageable tasks.

Step 4. Assess Risks: after carrying out hazard identification, either an existing risk assessment is used to see its suitability for the current task or a new one is conducted by a competent risk assessor. (See next section on risk assessment for the full process.)

Figure 2.24 Typical PTW process

Step 5. Review and Agree: once all documentation is compiled after gathering necessary information, discussions with team members and management, an agreement is obtained from all stakeholders. In order to indicate agreement, a signature must be obtained on the PTW.

Step 6. Employ Controls: prior to the issuance of PTW, controls must be put in place after assessing their availability and suitability. Upon confirmation of this step, the PTW can be issued.

Step 7. Issue PTW: finally, when all requirements have been met, PTW can be issued with several copies such as keeping a copy in the file, another copy in the permit office and one copy displayed at the location where the task is being undertaken.

2.6.2 Considerations when PTW is issued

1. PTW Display: an approved PTW must be displayed at all times at the location where the task is being undertaken or at a place close to the location of work. An additional copy should be placed in the control room.

2. PTW Suspension: situations where all PTWs will automatically be suspended include when a general alarm is raised and all personnel have to evacuate to the muster stations. It must be noted that a PTW remains live until cancelled even when it has been suspended. Other circumstances when a PTW may be suspended include:

   - when the task being undertaken conflicts with another task requiring PTW;
   - due to lack of required competent workers;
   - due to lack of tools or spare parts;
   - due to change in circumstances such as weather conditions, illumination or other controlling factors that will directly impact upon the task at hand;
   - when the nature of work changes due to interruptions or unseen work.

   A note stating the reason and duration for which a PTW was suspended must be entered on it. Any other comments that specify the condition of equipment and status of the task must also be noted. Once a PTW has been suspended, it cannot be reinstated until the issuing authority has reassessed the situation and has reapproved the PTW or issued a new one. In some cases, it may be prudent to cancel the suspended PTW and issue a new one to avoid complications and to ensure compliance with all safety requirements.

3. PTW Duration: duration of a permit depends upon the length of the task. It may be issued for a continuous period such as seven days provided suitable shift handover procedures and

revalidation requirements within this period are agreed. Many operators prefer the PTW to be cancelled at the end of each shift and a new one issued.

4. PTW Associated Procedures: most of the PTWs will have some associated procedures such as for isolation or overrides of equipment, pressure vessels or power supply, etc. Any such agreed procedures and other relevant information must be attached to the PTW to ensure all stakeholders are aware of the details.

5. PTW Cancellation: a PTW can be cancelled when the task has been completed and authorised personnel have certified it. Other occasions when a PTW will be cancelled are when:
   • its allocated time limit has expired;
   • there is a need to change the system of work;
   • the work has been stopped through SSOW intervention due to safety being compromised.

6. PTW Handback: when handing the PTW back to the issuing authority (originator), appropriate records should be entered in the system referring to the PTW number and stating the reason such as completion, cancellation or suspension because of which the PTW has been handed back. In the case of completion, someone must check the completed work for confirmation prior to filing the handed-back PTW.

7. PTW Colour Coding: this is only advisory to enable the colour of the permit document to specify the nature of task being undertaken. HSE recommends the colour coding[16] in Table 2.3 for different types of PTWs and associated procedures.

8. PTW Records: for completeness of the system and ensuring an appropriate audit trail, PTWs should be retained by the person responsible for maintaining records for a period of at least four years. For remote sites such as Normally Unmanned Installations (NUI), the permits must be kept on site for at least 30 days after completion of work.

Figure 2.25 In certain high hazard areas, only authorised workers should be permitted entry

Table 2.3 – PTW colour coding	
PTW or Certificate Type	**Colour**
Hot work	Red
Cold work	Blue
Confined space entry (PTW or other document such as gas entry certificate)	Green
Breaking containment certificate	Black edges
Equipment isolation certificate	White
High voltage isolation certificate	Yellow
Diving certification	White or other colour associated with diver's task

2.7 Risk Assessment

Risk assessment,[17] also referred to as Task Risk Assessment (TRA), is considered a very complicated process by many in the industry. In simple terms, every person carries out 'little' risk assessments at all times. For example, before travelling, a person will consider the prevalent weather. If the forecast is for rain, then an umbrella might be helpful to protect from getting wet – a potential harm. If the forecast is for heavy showers, then an additional rain coat/suit may be more appropriate to protect from the same harm. However, if the forecast is for a storm accompanied with thunder and lightning, then 'getting wet' may not be a major concern as compared to the 'fear' of being struck by lightning or of being blown away by strong winds – the harms may be 'death' or 'an injury'. Therefore the person may decide to stay indoors and wait for weather to improve – the only control measure being postponement of the task.

It can be seen from the foregoing analogy that as the severity of harm increases, the control measures (umbrella, rain coat or staying indoors) change. This is exactly the process that needs to be followed for all tasks. However, in order to ensure accurate workplace risk assessments are carried out, it is paramount to understand the basic terminology as explained here.

HARM can be defined as a physical injury to a person, damage to health, property or the environment.

HAZARD on the other hand is any unsafe act or condition, material, object or activity, circumstance or situation that has a potential for an undesirable adverse outcome such as damage to property, life or health of an employee or environment.

RISK refers to the combined outcome of the POSSIBILITY and SEVERITY of harm caused by a hazard or combination of hazards.

RISK ASSESSMENT is the study of hazards to ensure precautions (control measures) to reduce the risk of harm and where possible, prevent it. The most important factor for risk assessment is to consider each hazard and determine whether sufficient precautions have been taken to deal with associated risks. A risk assessment can be carried out by (see Figure 2.26):

1. Task selection.
2. Hazard Identification (HazID).
3. Consideration of the likelihood of occurrence of harm and severity of harm (if it can occur).
4. Determination of risk levels.
5. Establishing whether existing controls or precautions are sufficient.
   a. If sufficient, carry on with the task taking full notes to maintain records.
   b. If insufficient perform a Cost Benefit Analysis (CBA).

    i. If the costs are 'affordable i.e. low costs', add more controls and then repeat the cycle to re-evaluate risks until a satisfactory conclusion is reached.

    ii. If the costs are 'unaffordable i.e. high cost', postpone, delay or even cancel the task until such time that the costs become 'affordable'.

6. Check if a PTW is required to avoid conflict with other tasks.
7. Obtain approval from the final signatory such as area control or even the OIM.
8. Once approved, review with the work team and conduct toolbox talk.
9. When the work commences, a dynamic risk assessment will continue to be undertaken. If new hazards are identified, the task may need to be stopped until the resulting risks have been mitigated.

It is a widespread misconception in the industry that only the 'safety officer' or 'safety representative' is required to undertake risk assessments. It is rightly so as risk assessment and all associated tasks are part of their job. But as a general rule, every worker should be encouraged to carry out informal risk assessments and establish if any detailed risk assessment is required, then the safety officer should undertake detailed risk assessments to formalise the process and ensure all associated procedures are fully followed. The employer is obliged to assess risks for the following reasons:

1. to comply with national/international legislative requirements;
2. to reduce the losses and hence to improve profits;
3. to save life and property;
4. to protect the environment.

Whenever workers are assigned a job, they must first check if it is:

- An existing task – then they must check existing risk assessment and procedures to ensure compliance as well as understanding by all workers involved.
- A new task – the process described below should be followed for a complete risk assessment.
- A low risk task – the tasks can be carried out by a competent person who is familiar with the task and all associated requirements.

Once a 'new' task has been chosen, HazID is the most important process in a risk assessment. It is intended not only to identify hazards but also to create an awareness of hazards, and to identify the personnel who are at risk. The process of HazID is discussed in detail in the next subsection.

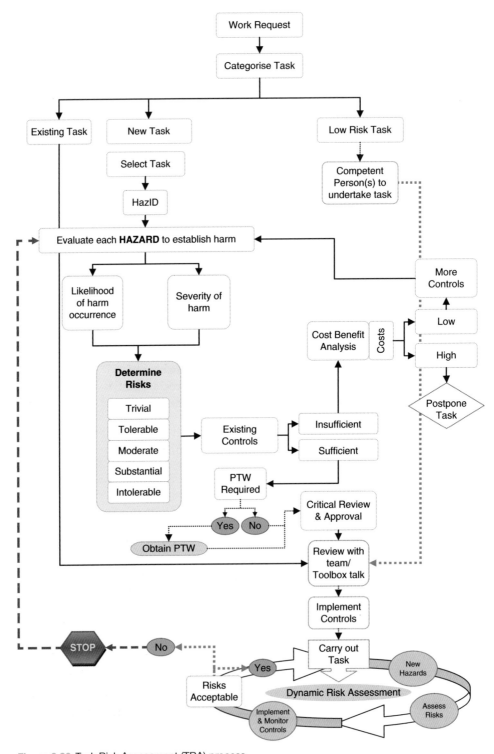

Figure 2.26 Task Risk Assessment (TRA) process

2.7.1 Task Selection

It is always recommended to break a larger job or project into smaller tasks so that each task, associated hazards and risks can be assessed separately. This provides small, easily manageable tasks which can be looked after without problems. For example, if it is intended to inspect a faulty 'gate valve' in a ballast tank, it is prudent to divide the task into two smaller tasks:

- entry into enclosed space;
- inspect the valve and carry out repairs if required.

Most safety management systems provide a generic risk assessment for the first task i.e. entry into enclosed space. The second task of valve inspection will decide the subsequent tasks that may have to be performed. It may only require clearing the valve seat, but for that at least one flange will require to be opened. In ballast tanks, it is not uncommon for the nuts and bolts to get rusty, which may require the use of a grinder or oxyacetylene cutting to remove them. If that is the case, then a 'hot work permit' and a separate risk assessment will be required. In any case, if available, the generic risk assessments should be used as a guide to assess if any of the circumstances have changed as they may vary the hazards or their nature. Once it has been established that there has been no change in circumstances, only then should a generic risk assessment be used.

2.7.2 Hazard Identification (HazID)

A hazard can be defined as any unsafe act or condition, material, object or activity, circumstance or situation that has a potential for an undesirable adverse outcome such as damage to property, environment, life or health of an employee. The purpose of any risk assessment is to consider the probability of events or outcomes that can be seen as adverse, harmful or just undesired. However, none of these can be considered unless the hazard has been identified. Therefore the process of risk assessment begins with 'studying' the hazards so as to be in control of any potential adverse outcomes. The risk assessor must keep in mind that this phase neither provides an indication of the magnitude of the loss or damage nor the likelihood of such an event taking place. At the same time, they must also remember that the subsequent risk analysis can be based on the findings during the 'hazard identification'.

In short, the purpose of HazID is to systematically study the processes, practices, procedures and actions so as to identify the causes and consequences of deviations from the intended objective with an overall goal that each possible hazard can be recognised, recorded and measures taken to prevent the consequential loss or damage.

Sources of hazards can be considered under four categories i.e. plant, place, competency and system[18] as given below:

Plant:

- falling objects;
- dropping objects;
- sharp objects;
- noise;
- slippery surfaces;
- chemical reaction;
- poisoning;
- pollution;
- electrical hazards;
- moving objects, machinery or tools;
- harmful dust or rust when preparing steel surface for paintwork;
- sudden movements including rolling, pitching of the installation or movement of other objects associated with it.

Place:

- workplace layout;
- obstructions in the work area such as location of objects and other workers within;
- extreme temperatures;
- exposure to chemicals such as solvents or paints;
- illumination;
- means of access or egress.

Competence:

- interaction amongst workers, helping and cooperating with others in contrast with lack of help and support;
- inadequate knowledge or training;
- unsafe acts or behaviour;
- lack of attention or concentration;
- inability to perform tasks;
- errors and omissions;
- fatigue;
- influence of medication or other physical limitations.

System:

- stress or fatigue;
- lack of guidance or monitoring;
- inadequacy or non-provision of tools, PPE/RPE or other resources;
- insufficient rewards and support;
- ineffective communication.

The HazID techniques can be divided into four categories[19] based on the area in which they are applied or used:

1. *Process hazards.* The hazards that may occur due to processes or procedures adopted as a deviation from original action can

Figure 2.27 Type of task requiring risk assessment
(Source: © Petrofac Training)

be identified and evaluated. The basic concept is to look for any possibilities of unwanted 'developments' or an 'undesired way' in which a task may progress. A useful technique to identify hazards is to carry out a 'SWOT analysis' (discussed later in this chapter). The examples of such processes can include change in temperature, weather (rain, wind, etc.) or non-completion of the job in the planned allocated time that may cause excessive fatigue to workers.

2. *Hardware hazards.* The hazards that may occur due to failure or insufficiency of tools or equipment (plant) are considered in this category. When performing hardware hazard identification, it is important to consider the 'consequences' of failure of any tool or equipment so that control measures can be put in place to avoid any mishaps. In the example of replacing a valve given

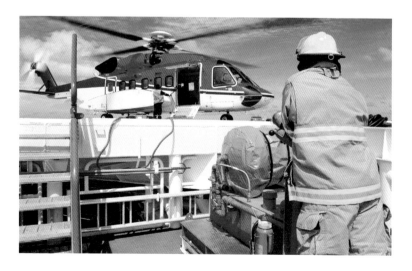

Figure 2.28 HLO and the helideck team must always be ready to deal with any emergencies.

earlier, if oxyacetylene is used, a need to use an 'exhaust' fan continuously should be one of the control measures. Failure of the exhaust fan can be identified as a hardware hazard which must be dealt with by placing control measures i.e. a standby exhaust fan, use of SCBA or in the worst case, deferring the task till an alternative can be provided.

3. *Control hazards*. This category of hazard identification primarily applies to hazards that may develop as a result of failure of 'computerised' control systems. For on-board offshore installations or vessels, where a computerised control system is used for operations such as ballasting/deballasting, liquid cargo operations or any other operations, the risk assessor should make sure that failure of control system has been considered and manual controls located, procedures set in place and personnel practised to tackle these situations. The means to tackle failure of computerised controls is a 'control measure' for the risks associated with each operation.

    The same category of hazard identification can also be used for a 'system' that controls various processes or tasks. For example for entry into enclosed space, the personnel who enter the enclosed space are required to maintain contact (e.g. by portable radio) with a person supervising the task from outside. If the supervisor becomes unavailable due to any mishap (usually not considered as a hazard), then there should be some procedure in place to make other decisions before the hazard can lead to a harm.

4. *Human hazards*. These hazards are directly or indirectly caused due to involvement of personnel in almost all processes. These are generally termed as the 'human factor' which tends to be always a contributory factor in accidents. The human hazard can be their error, behaviour or competency. These hazards should consider interaction of humans (workers) with the above three hazard categories i.e. process, hardware and control systems. For example non-cooperation or lack of understanding between two co-workers who are assigned the same task should be treated as a hazard. The possibility of error when assigning a sensitive task to a trainee as compared to an experienced person should also be considered to avoid any accident.

2.7.3 Hazard Monitoring

Hazard monitoring essentially means managing all identified risks. For example, lack of oxygen in an enclosed space is a hazard; the person entering this space with an SCBA is provided with a means to monitor the air pressure in the air bottle by a low pressure alarm. If this alarm does not operate, then the hazard monitoring does not exist and hence the risks may be the same as going in the compartment without SCBA. As a back-up therefore, the supervisor should note the time of entry into the space along with the time lapsed for which the air has been used.

Hazard monitoring can be divided into two types:[20]

* active monitoring to achieve a safety culture by inspections, surveillance, auditing and reviewing to ensure a 'control' on hazards and minimisation of 'incidents';
* reactive monitoring as a consequence of any failures resulting in incidents.

2.7.4 Cost Benefit Analysis (CBA)

CBA is primarily a business tool but it can be used for any task where an outcome of a planned action needs to be assessed. In business terms, it is used to calculate net profits. However, when used for risk assessment, the costs can include time, number of personnel (man hours), equipment, tools and other consumables. An example of a consumable for entry into an enclosed space (with lack of oxygen) to replace a damaged valve can be the air bottles of SCBA amongst other consumables. The benefits may be the priority for completion of the job e.g. the job may be critical for the safety of the structure or operations. As such, scores cannot be given to either cost or benefits for a comparison. The deciding factor is the job priority e.g. in the previous example of replacing the valve, consider the consequences of delay in repairing it. If it is a critical valve without which the integrity or safety of the installation may be compromised, then the task will need to be completed even if the costs are high. Nevertheless the harms should be evaluated, the risks should be estimated and additional control measures placed so as to minimise the risks.

2.7.5 SWOT Analysis

SWOT (Strengths, Weaknesses, Opportunities and Threats) analysis is a basic tool used for audits and development of marketing plans. However, this tool can be used by safety officers to analyse any hazards that may pose a risk. Whether it is dealing with a single person's behaviour that is causing concerns or an accomplishment of a new task which may need some consideration for planning, the safety officer can use the four elements i.e. Strengths, Weaknesses, Opportunities and Threats to increase the possibilities of successful task accomplishment.

The basic idea for the use of SWOT analysis is to develop a plan to consider internal factors (Strengths and Weaknesses) and external factors (Opportunities and Threats) to

* make the best use of strengths;
* explore the highest number of opportunities;

Figure 2.29 Hazard identification must be from the perspective of workers

- minimise, overcome or reduce the impact of weaknesses; and
- limit any threats.

2.7.6 Evaluating Hazards and Assessing Risks

Once the hazards have been identified, they need to be evaluated to find risks by checking the following two items in parallel but independent of each other:

- likelihood of harm occurrence;
- severity of harm if it occurs due to a particular hazard.

Each identified hazard can be evaluated and associated risks assessed by considering the guidance given below:

- Collect information about hazards from sources such as the personnel who are aware of any past incidents or from suppliers, manufactures or from MSDS (Material Safety Data Sheets) provided by suppliers or other documentation that may be available.
- Obtain information from guidance given in the ACoP (Approved Codes of Practice), Safety Management System and other national or international regulations.
- Information from previous accident/incident reports or risk assessments should be used for evaluating hazards. It must be noted that if a large number of hazards are identified, not all of them will be 'serious' or 'threatening'. Any hazards that are considered 'trivial' or 'inconsequential' can be ignored for detailed evaluation once it has been established that they will not cause any harm.
- Once the hazards have been selected for further investigations, then a priority list needs to be prepared in which the harms are 'ranked' or 'classified' according to the likelihood of their occurrence and severity of the harm.
- The standard ranking for harms used in many industries is shown here for ease of understanding. There are other ranking systems that may be used for both likelihood and severity of harm. The scores are given by numbers as shown in Table 2.4 to resolve ambiguity arising from the use of words for likelihood or severity of harm.
- When rating likelihood, factors such as the number of personnel exposed to the harm, the time for which they are exposed and the nature of the hazard should be considered. Any score found is not absolute as the control measures may fail or the circumstances may change. Rating likelihood of occurrence of a harm does not mean to rate the 'magnitude of injury' but it is to rate the likelihood that an injury will occur. For example, people have fallen when working aloft with no severe injuries but at the

same time there have been incidents where people just tripped over in the dining room, smashed their head on the table and had a serious head injury. This example is particularly significant to differentiate the circumstances as there may be no 'risk assessment' for personnel going to the dining room for dinner.

• When rating severity, it is important to compare and contrast various situations to see how severe the injury would be if an accident were to happen. For example, a person hit by a falling object without any hard hat will sustain more severe injury as compared to a person with a hard hat on.

• The risk level or score is estimated by formula.

$$\text{Risk Level} = \text{Likelihood of Harm} \times \text{Severity of Harm}$$

The action and controls required for the estimated level or score of risk is taken according to the guidelines given in Table 2.5.

• Most organisations have their own 'scoring' or 'risk level' estimation systems. Many use a score of up to '4' for both likelihood of harm and severity of harm. Regardless of the scoring mechanism, as long as the risks have been estimated and controls placed, the numbers do not really matter. However, when the risk levels are estimated, the highest levels should be considered for ensuring optimum safety.

• For each risk assessment, the process of identifying and evaluating the hazard to find the risk level may have to be repeated a number of times until it has been effectively established either:
  • that the controls are sufficient and the task can go ahead (a final assessment will reveal the 'residual risks'); or
  • the sufficient controls cannot be achieved due to high cost and the task has to be postponed.

• A risk assessment carried out with a POSITIVE identification of control measures for all hazards will ONLY be valid until the identified controls are IN PLACE. The moment a control is 'mislaid', the risk assessment becomes VOID unless the controls are brought back in place. It means that the risk assessor has to review and monitor the assessments on a continual basis.

2.7.7 Controlling the Risks and Action Plan

Whilst one purpose of risk assessments is to establish the level of risk, the second purpose is also to find, place and ensure to keep in place the controls that may be required and are identified during the risk assessment process. In practice, the control measures can be anything that reduces risk level by either reducing the likelihood of occurrence or severity of harm. There are two reasons for controlling risks:

• To comply with legislative requirements.
• If the legislation does not provide clear and specific guidance,

Table 2.4 – Ranking and scoring of harms

	Ranking	Score	Notes and Examples
Likelihood of Harm	Highly Unlikely / Improbable	1	The harm unlikely to ever occur to a competent person i.e. the harm hardly ever occurs. *Example:* The likelihood that a competent person working aloft on a bosun's chair, properly attired with PPE including safety harness in *light breeze* will fall down is 'highly unlikely' hence a score of 1 can be given.
	Unlikely / Remote	2	The harm likely to occur ONLY ONCE to a competent person, consequently the person will 'learn from mistake' to avoid reoccurrence. *Example:* The likelihood that a competent person working aloft on a bosun's chair, properly attired with PPE including safety harness will fall down *in strong wind* is 'unlikely' as they have the safety harness that will prevent them from falling down even if they slip, trip or are pushed away due to wind; hence a score of 2 can be given.
	Likely / Probable	3	The harm may occur EVEN to a competent person i.e. it is certain that the harm will occur. *Example:* The likelihood that a competent person working aloft on a bosun's chair, properly attired with PPE with *safety harness NOT secured to a strong point* will fall down *in strong wind* is 'likely' if they slip, trip or are pushed away due to wind; hence the highest score of 3 can be given.
Severity of Harm	Slightly Harmful / Negligible	1	The harm can cause ill-health associated with temporary discomfort such as pain, allergy, headache or other external injuries. *Example:* Personnel working on deck in the presence of smoke which may cause discomfort even if masks are being used. As the severity of harm does not cause a long-term ailment, the score is 1.
	Harmful / Serious	2	The harm can cause ill-health which may cause minor permanent damage but the consequences are not serious e.g. burns, cuts and bruises that will leave a permanent scar or short-term disability. *Example:* Someone working in an enclosed space with a continuous intake of fresh air by blower fans where hot work is also being carried out, the severity of harm would be serious such as suffocation or loss of consciousness if the blower fans fail, hence the severity of harm should be given an increased score i.e. 2 even though the likelihood may have a low score (1) as it will happen only IF the fans fail.
	Extremely Harmful / Calamitous	3	The harm can be imminent and effect on life of a person can be illnesses or injuries that may render the person disabled for a long time e.g. a fracture or cancer etc. *Example:* A person falling from a height when working aloft should be given a high score i.e. 3 for severity of harm as the consequences of a fall would be calamitous such as fracture, excessive bleeding or even instant death. However, the likelihood may have a low score (1) due to other control measures.

Table 2.5 – Risk levels and scores

			Severity of Harm					
			Slightly Harmful 1		**Harmful** 2		**Extremely Harmful** 3	
Likelihood of Harm	**Highly Unlikely**	1	Trivial	$1 \times 1 = 1$	Tolerable	$1 \times 2 = 2$	Moderate	$1 \times 3 = 3$
	Unlikely	2	Tolerable	$2 \times 1 = 2$	Moderate	$2 \times 2 = 4$	Substantial	$2 \times 3 = 6$
	Likely	3	Moderate	$3 \times 1 = 3$	Substantial	$3 \times 2 = 6$	Intolerable	$3 \times 3 = 9$
			Risk Level	**Score**	**Risk Level**	**Score**	**Risk Level**	Score

Table 2.6 – Risk levels and actions

Risk Level	Risk Score	Actions and Controls Required
Trivial Risks	1	Risks that can be deemed unimportant and so additional action to reduce such risks is not normally required
Tolerable Risks	2	Risks that can be tolerated/accepted without any possible harm but monitoring is required to maintain adequacy of existing controls. CBA may provide an opportunity for alternative solutions.
Moderate Risks	3–4	Risks that require additional resources to achieve substantial control of potential harm, with a possibility of slight increase in the cost. The allocation of resources should be considered on the basis of severity of harm e.g. if it is Moderate, then a further assessment may be required to either further subdivide the job into smaller easily manageable tasks or controls may need to be increased.
Substantial Risks	5–6	Risks that are unacceptable and require effort to reduce them at any cost. In this case the balance between the cost and benefits may incline towards more cost to achieve the same benefits but will ensure safety.
Intolerable Risks	6–9	Risks that are unacceptable because of the level of severity, unavailability of resources to allocate such that risks can be neither controlled nor reduced or the cost is too high to gain any benefits. In this case, the task cannot continue.

to show 'due diligence' for protection of personnel. Employers and employees are both required to show due diligence for health and safety of workers. It means that each person is required to make efforts to avoid harm or injury to themself and others by adequate levels of judgement, forethought, care, determination and competency under the given circumstances. Violation of due diligence as required under legislation is considered as an act of 'negligence' which goes against the requirements of the HASAWA and, if proven, may be deemed a criminal offence.

Selecting controls for risk mitigation is not an easy task. The hierarchy of controls[21] given below can be used to formulate an action plan that can ensure hazard control:

1. ELIMINATE the hazard at its source to circumvent risks. This is the most effective way of controlling hazards as the hazard is not present so there is no possibility of harm.
2. ENGINEER CONTROLS by
   * SUBSTITUTION for less hazardous material or process e.g. a hazardous chemical may be replaced by a less hazardous chemical or part of a job may be postponed until circumstances change.
   * Ensuring work ERGONOMY – i.e. adapt the work to the worker not the worker to the work e.g. tools or material can be identified that can help ease the task.
   * Use of BARRIERS to control exposure to substance, material or process e.g. enclosure of substances or isolation of workers can act as barriers that will remove the possibility of contact between the substance and the worker; hence chances of harm will be controlled.
3. ESTABLISH CONTROLS such as by Warning, Instructing and Training personnel for tools, equipment or other measures to prevent harm from the identified hazards.
4. EVALUATE risks that cannot be avoided.
5. PPE – These should be used as a last resort. For example, a hard hat is meant to protect the wearer from head injury caused by a falling object. It may not provide protection if the wearer knowingly allows the object to fall and deliberately places their head under the falling object.
6. EMERGENCY PREPAREDNESS – The personnel involved in a task where the likelihood of occurrence of harm or its severity has been reduced by control measures should also be advised and trained on all cases when the controls fail and they have to take evasive actions to avoid injury.

2.7.8 Documentation for Risk Assessment

The risk assessment documentation is important because it can be used for:

1. effective communication between all concerned parties;
2. collecting information during risk assessment process;
3. supplying information for further investigations, risk assessments and monitoring;
4. allowing discussion and feedback, particularly in safety committee meetings;
5. reference, proof and justification of actions and decisions – particularly if an accident happens and investigations require officials to submit documentation for arbitration or court cases.

Due to the above reasons, risk assessment documentation must at least contain the following items:

Figure 2.30 All significant risks must be documented

1. *General Information*:
   a. Title of document, reference number, date and time to differentiate each document from others. Generally it is the responsibility of the management to devise a system for maintaining records and educating safety officials including risk assessors and workers at the receiving end of instruction.
   b. If a generic risk assessment is used, the actual reference number and the version or revision number (if applicable). For example, a generic entry into enclosed space risk assessment may have three versions i.e. Version 1 for entry into ballast tank, Version 2 for entry into a fresh water tank and Version 3 for entry into a fuel oil tank. These may be further revised if the circumstances change.
   c. Name and designation of personnel who conducted the risk assessment.
   d. Identification of the installation or vessel and other factors such as weather at the time when the assessment was carried out.
   e. Record of approval including the person who approved it along with date and time of approval.
2. *Hazard Specific Information*:[22]
   a. The selected task i.e. the name of the task for which the risk assessment is carried out.
   b. The identified hazards.
   c. Details of entities (workers, processes, equipment, property or environment) exposed to hazard.
   d. The evaluation of hazards including the scores given to likelihood of harm and its severity.
   e. The estimated risk levels.
   f. The controls in place for all identified hazards.
   g. The monitoring procedures or instructions.

2.7.9 Risk Assessment – The Simplified Steps

Step 1. Select a task.
Step 2. Divide the task to smaller easily manageable tasks.
Step 3. Compare the present task with the ones in existing or generic risk assessments so that the full process does not have to be repeated. If a 'similar' risk assessment can be found, study it to ensure the 'calculated risk factors' are for exactly the same circumstances and conditions, if not carry out a new assessment by continuing to Step 4.
Step 4. Study each task to identify hazards and personnel (or other entities) who may be exposed to the identified hazards.
Step 5. Evaluate each hazard; discard the ones which have no potential for harm, select the ones that are serious.

Step 6. Rate each serious hazard for likelihood of occurrence of harm.

Step 7. Rate each serious hazard for severity of harm.

Step 8. Multiply likelihood by severity to estimate risk level for each hazard.

Step 9. Locate and study possible control measures (from hierarchy of controls) to reduce either the likelihood or severity of harm.

Step 10. Repeat Steps 6 and 7 to find the new rating or score.

Step 11. Repeat Step 8 to find new level of risk.

Step 12.
- If level of risk 'Substantial (Score 5–6 or less)', then the controls are sufficient and the task can go ahead;
- If level of risk 'Intolerable (Score 6–9)' then place additional controls; however, if the cost is too high then the task has to be postponed or an alternative solution sought.

Step 13. Assess residual risks.

Step 14. Check if a PTW is required. If required, then follow the procedures for obtaining the PTW.

Step 15. Document all findings whether it is decided to proceed with the task or to postpone it.

Step 16. If it is decided to proceed with the task, monitor continuously to keep controls in place. Additional steps such as toolbox talk, implementation of controls should also be considered at this stage.

Step 17. When the task is being undertaken, a dynamic risk assessment for any change in circumstances must continue throughout the life of the task. If at any stage new hazards are identified, further risk assessment must be carried out as stated above and shown in the flow chart in Figure 2.26.

2.7.10 Sample Risk Assessment

The risk assessment in Table 2.7 is for entry into a ballast tank for routine inspection. Note that a scoring of 3 (as discussed in this chapter) for likelihood of occurrence and severity of harm has been used.

2.8 Toolbox Talks

A toolbox talk is a short 'safety meeting' about the task being undertaken to ensure accurate and full information about procedures as well as any hazards, risks and control measures is exchanged between the person in charge of the task and the workers.

Historically, toolbox talks were used by construction workers as a documented safety meeting where the attendees stood around the toolbox before they started the work, hence the name toolbox talk.

Table 2.7 – Risk levels and scores

Sample 2 – Entry into Enclosed Space Risk Assessment – Ballast Tank	Rig1

Risk Assessment Number ………………………	*Relevant Generic RA No. ………………………*
Assessed by: ……………………………………………. *Date: …………………………………………………………*	*Approved By: ……………………………………………* *Date: ……………………………………………………………*

Hazard(s):

Entities Exposed

H1. Lack of oxygen

H2. Darkness

H3. Slippery surface

All personnel entering the space

Risk(s)/Harm(s):

R1. Unconsciousness.

R2. Tripping from ladders and /or other structure within tank

R3. same as R2

Initial Risk Assessment:

Likelihood of Harm				Severity of Harm			Risk Level
☑	3	Likely	x	☑	3	Extremely Harmful	3 x 3 = 9
	2	Unlikely			2	Harmful	
	1	Highly Unlikely			1	Slightly Harmful	

Outcome of Initial Risk Assessment:

Risk Level 9 – Intolerable Risk (see table below): The task of 'entering into enclosed space' cannot continue until controls are placed/increased to reduce likelihood and/or severity of harm.

Control Measures:

Who is to Monitor:

For H1R1. Use oxygen analyser to test atmosphere and continuous ventilation to replace tank atmosphere with fresh air. Test again prior to entry

- Team leader

(continued)

Table 2.7 – (continued)

For H2R2.	Use portable lights as a primary means of illumination and portable emergency torch lights as a secondary means of illumination	• Welder
For H3R3.	Use PPE such as leather hand gloves, hard hat and safety shoes	• All personnel involved with the task

Re-assessment of Risks with Control Measures:

Likelihood of Harm			Severity of Harm			Risk Level
	3	Likely	x ☑	3	Extremely Harmful	1 x 3 = 3
	2	Unlikely		2	Harmful	
☑	1	Highly Unlikely		1	Slightly Harmful	

Outcome of Re-Assessment:

By placing the control measures given above, the likelihood of harm has been reduced from a rating of 3 to rating of 1 which reduces the risk level from 9 to 3. It means that the risks are still 'moderate' for which additional controls such as oxygen analyser, blower fans, cargo lights are required to ensure safety.

Risk Level	Risk Score	Actions and Controls Required
Trivial Risks	1	Additional action to reduce such risks is not normally required
Tolerable Risks	2	Those risks that can be tolerated/accepted without any possible harm but monitoring is required to maintain existing controls.
Moderate Risks	3–4	Additional resources are required to achieve substantial control of potential risk, with a possibility of slight increase in cost.
Substantial Risks	5–6	Those risks that are unacceptable and require effort to reduce them at any cost.
Intolerable Risks	6–9	Those risks that are unacceptable because of the level of the severity, unavailability of resources to allocate such that risks can be neither controlled nor reduced or the cost is too high to gain any benefits. In this case, the task cannot continue.

Figure 2.31 Toolbox Talk – a final opportunity to discuss before undertaking the task

However, this practice proved so useful that most industries have now adopted it to suit their needs.

Ideally, the toolbox talks should be held between the team members deployed on any task immediately prior to every new task and at the beginning of every shift. Further toolbox talks should be held as soon as anything changes e.g. work environment, equipment or personnel to ensure any changes in procedures are adopted accordingly and new risks are considered for appropriate control measures.

Toolbox talks form part of many safety management systems which require these talks to be formally recorded, mostly on a special form with a copy held locally and another sent to senior management. These records will form part of documents that will be checked during both internal and external audits. Generally, the following items are covered and recorded for each toolbox talk:

* name of the person in charge of the talk;
* name of the main project/task;
* names of attendees including signatures of all attendees;
* topics discussed including reference to any operating procedures, risk assessments, PTW, etc.;
* issues identified and solutions proposed and agreed;
* any unresolved issues referred to senior management;
* good practices shared, training or instructions provided;
* date and time (duration) of the talk.

2.9 Accident Reporting Procedures

The offshore industry, after years of hard work, has now established a 'non-blame culture' for improvement of safety. Therefore each individual must report all incidents, however trivial, without any fear of penalty.

Figure 2.32 All accidents must be reported according to agreed procedures

Upon joining, workers must establish the relevant procedures and know the line of reporting for incidents or accidents. Generally, workers will report incidents to their immediate supervisor or safety officer. In all cases of a serious accident where a grave and imminent danger to life, environment or property exists, an appropriate alarm must be raised and the control room informed immediately. If the accident involves an injury, then the medic must also be informed and given the initial information.

In the event of an incident or accident, the offshore operator/duty holder will instigate an internal investigation. In addition, under the Reporting of Injuries, Diseases and Dangerous Occurrences Regulations 1995, certain incidents or accidents must be reported to the HSE. This report will be sent on a pre-specified form by the OIM. Depending upon the severity of the accident, HSE may send their inspectors to investigate the accident where evidence from the place of accident, witness statements from those who saw the accident and from others directly or indirectly involved in that particular area may also be required.

2.10 Safety Committees and Meetings

Safety committees, representatives and constituencies on-board offshore installations are managed under the Offshore Installations (Safety Representatives and Safety Committees) Regulations 1989 (SI 1989 No. 971) in the UK. Various requirements under these regulations are described here.

2.10.1 Safety Constituencies

Safety constituencies are subdivisions of the offshore installation to ensure safety of a group of workers is coordinated by a safety representative who can represent members' views to the safety committee. Every installation is required to establish safety constituencies on the basis of areas of the installation, activities undertaken, nature of hazards and other criteria that may require the duty holder to consider further requirements specific to the installation. The allocation of workforce is based on the factors below:

Figure 2.33 Safety officials must implement agreed procedures

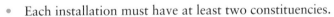

- Each installation must have at least two constituencies.
- Every existing or new worker shall be assigned to one constituency. The new workers are required to be given this information in writing.
- The minimum number of workers assigned to any constituency shall not be less than three and not exceed 40.
- Workers shall also be informed about their safety representative.
- Duty holders shall inform the safety representative, in writing, of any new members to their constituency.

- The allocation of constituency is not required for any personnel on-board installations whose stay on-board is not expected to be more than 48 hours.

When constituencies are established, the duty holder is required to display their layout at prominent places within the installation. If the workforce consists of speakers of languages other than English, then that language should also be used to convey information to those personnel.

2.10.2 Safety Representative

According to the Offshore Installations (Safety Representatives and Safety Committees) Regulations 1989, the offshore workers are entitled to elect or nominate a safety representative. The election is held through nominations by the members:

- when a constituency is established or modified;
- after two years since the last election;
- when the existing safety representative ceases to be a safety representative or if their job is terminated through resignation or otherwise.

When the elections are to be held, the duty holder will advertise the details of the constituency in which elections are to be held including the names of members and the date on which this election will be held. After the given period for nominations, the list of nominees will be displayed at prominent locations within the installation.

Figure 2.34 Safety representatives can inspect workplaces for hazards

If more than one candidate is nominated, then a secret ballot will take place. Results of the ballot will be published in no more than a week after the ballot. In the case where no candidates wish to be nominated, the duty holder will hold this position and continue to follow the above procedure for electing a safety representative.

According to the Offshore Installations (Safety Representatives and Safety Committees) Regulations 1989, safety representatives have functions and powers to fulfil their functions but no legal duties.

The functions of a safety representative are:

1. To investigate potential hazards and dangerous occurrences and to examine the causes of accidents in constituencies where members of their constituency may be involved or affected. This function may be extended to other constituencies where a safety representative is not available.
2. To investigate complaints by any member of their constituency relating to occupational health and safety.
3. To make representations to the duty holder(s) for matters related to health and safety or investigations as described in 1 or 2 above.
4. To attend meetings of the safety committee.
5. To represent their constituency members' issues in internal or external consultations.
6. To consult members of their constituency either individually or collectively on any matters related to hazards, dangerous occurrences, accident investigations or other health and safety matters.

The powers of a safety representative are as follows:

1. The safety representatives may seek advice and guidance from persons on the offshore installation or elsewhere on any matters related to health and safety or accident investigations.
2. A safety representative may inspect any part of the offshore installation or its equipment if:
   a. they have given the installation manager and, if their employer is not the installation owner, their employer, reasonable notice in writing of their intention to do so; and
   b. they have not inspected that part of the installation or its equipment in the previous three months; and
   c. they may carry out more frequent inspections by agreement with the installation manager or their employer.
3. After notifying relevant personnel such as the duty holder or their employer, the safety representative may inspect part of the installation or equipment to determine the cause of an accident or incident when:

a. there has been a notifiable incident; and
b. it is safe for an inspection to be carried out; and
c. the interests of the members of their constituency might be involved.

A notifiable incident means any casualty, accident, injury or disease which is required to be notified to the HSE by the installation owner or the installation manager.

4. Where two or more safety representatives consider there is an imminent risk of serious personal injury arising from an activity carried out on the installation:

   a. they shall make representations to the installation manager who shall prepare and send a report in writing on the matter to an inspector appointed as soon as is reasonably practicable; and
   b. a safety representative may make a report in writing by the fastest practicable means to an inspector appointed.

5. A safety representative may receive information from the findings of investigations carried out by the inspectors.

In addition to the above powers, a safety representative is also entitled to:

- see and be supplied with copies of any document relating to health and safety of workers except the individually identifiable personal health records for workers;
- take time off without loss of pay to perform their functions as a safety representative and undertake training relevant to the role.

2.10.3 Safety Committees

The owner of an offshore installation (duty holder) is required to establish a safety committee where one or more safety representatives have been elected. Each committee shall include the installation manager as the chairperson along with another person appointed by the duty holder, all safety representatives and any other persons may be co-opted by committee members.

The first meeting of the committee must be called by the chairperson within six weeks of the establishment and, subsequently, one meeting every three months. When safety representatives are unable to attend any meetings, they can nominate another member from their constituency. The minimum number of members for a safety committee meeting is the chairperson and at least one-third of the safety representatives holding office when the meeting is called.

A safety committee has the following functions but not the duties:

1. to keep under review:
   a. the measures taken to ensure occupational health and safety of the workforce;

    b. recommendations made to the installation manager with regard to the system of the constituencies so as to ensure adequate representation;

    c. arrangements for the training of safety representatives;

    d. frequency of safety committee meetings, the circumstances under which they may be called to make representations to the installation manager or employer as appropriate;

2. to consider representations from any member of the safety committee on any matter affecting the occupational health and safety of the workforce and make recommendations to the installation manager as appropriate;

3. to consider the causes of accidents, dangerous occurrences and cases of occupational ill-health and make recommendations to the installation manager as appropriate;

4. to consider any document relating to the occupational health and safety of the workforce which is required by any statutory provision to be kept on the offshore installation, except any individual's personal health record;

5. to prepare and maintain a record of its business (meeting minutes). A copy of safety committee meeting minutes shall be kept on the installation for one year from the date of the meeting. A copy of these minutes shall be sent to the installation owner who shall keep it at a place onshore in the United Kingdom until the sixth anniversary of the expiry of the year to which the record relates.

Figure 2.35 Components of change

2.11 Management of Change

Constant innovation in technology and findings of accident/incident investigations drive a continual change in the offshore work environment. Automation in mechanical work, speed of communication and transportation dictate a constant need to update knowledge and therefore procedures. Ease of accessibility of information drives stakeholders to scrutinise each task, associated procedures and potential outcomes. However, when something needs to change, the workforce must be ready to accept the change prior to it being 'enforced' upon them. Like in any other industry, offshore industry also requires a system to manage change.

Management of change can be defined[23] as a combination of policies and procedures to evaluate the impact of proposed change in order to avoid it resulting in unacceptable risks that may lead to serious accidents. In order to ensure its success, the following steps should be considered to manage any change:

1. Recognise the change – various factors can affect the existing systems, procedures and equipment that may dictate the need for a change. This may even be suggested by the workforce who

can be directly or indirectly affected by the change or a lack of a change. However, not every change has to be managed. For example, replacing a component within existing machinery is not a change but when the whole machine is being replaced with a newer, modern and more automated make and model, then it can be considered as a change and must be managed.

2. Reviews
   a. Initial review – at this stage checks are made to see if the change involves replacement of a component or process. If it complies with specifications of the existing component or process on the basis of 'like for like' change; then it is considered as 'replacement in kind' and does not require management of change.
   b. Further review – if the initial review establishes justification for a need for change, particularly for health and safety reasons, then a further review is required to assess the overall impact and carry out a risk assessment.

3. Risk assessment – in some cases where no significant hazards are identified, there may not be any need for a risk assessment. However, in most cases, a risk assessment will be carried out by a competent person. This step will not only determine the implications on the safety of workers but will also consider an overall impact on outputs from business perspective. The findings will result in development of an implementation plan.

4. Approval – at this stage, an approval will be sought from all stakeholders by providing them all the information gathered in the previous stages. If an agreement is reached, the process can proceed to the next stage i.e. the implementation plan will be agreed.

5. Implementation – all tasks will be carried out as per the agreed implementation plan. This will encapsulate implementation of all changes whether they are procedures, equipment, personnel or their training for the new procedures or equipment. Even at this stage, the process may require reviewing each step to ensure conformity between the planned and achieved objectives. Any differences may require backtracking to the previous steps and agreeing any further changes.

6. Monitoring – once implemented, 'the change' processes will need to be monitored to ensure compliance at all times. Depending upon circumstances, the duration of monitoring may be determined by mutual agreement. Personnel involved must not hesitate to raise concerns as soon as any variations with a potential to cause harm are identified.

Whilst the above procedure can be put in place at any workplace, the most difficult phase in any change is to manage change in human behaviour. A graphical representation of change in behaviour with time is given in Figure 2.36. As it can be seen, a change will never be in a straight line but instead it moves through four phases:

1. Denial
2. Resistance
3. Exploration
4. Commitment.

Depending upon the nature of change and who initiated it, acceptance of change will depend upon workers' involvement from the outset as well as the impact any change will have on their work-life cycle. In order for workers to accept a full commitment, the following factors must be considered:

* Change must be implemented progressively to avoid a single event likely to affect a majority of the stakeholders.
* Each individual will naturally react in a slightly different way to the change request, hence a suitable time period must be allowed for acceptance of a full change. This highlights the need to continuously monitor even when a change has been fully implemented.
* Change presents opportunities for growth as well as some potential for taking risks, especially where some individuals continue to follow existing procedures despite the implementation of change. In order to ensure success, opportunities for training individuals to up-skill any skill gaps must be fulfilled in a timely fashion.
* Those who are tasked to monitor implementation of change must consider personnel who are consistently falling short of its acceptance or accepting it only because they have no choice. These individuals are most prone to swing in the backwards direction instead of moving swiftly though the four phases shown in Figure 2.36.

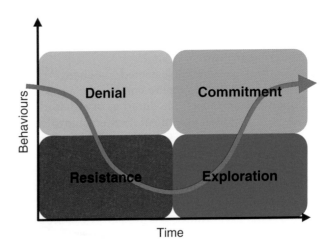

Figure 2.36 Management of change

2.12 Shift Handover Procedures

Many accidents have occurred because the full information about the tasks being undertaken, especially maintenance work, was not passed to the relief shift. In cases where workers go on short breaks and resume work where they left, communication is not an issue, but where the tasks need to continue even when they are on break or when they change over shift and hand over all 'running' tasks to the next shift, a simple breakdown or lack of communication can prove to be disastrous. The Piper Alpha disaster is a well-known example of such a failure where the relief shift was not informed about an incomplete task. Another accident that highlighted a failure of communication between the workers on two shifts was the Sellafield Beach incident in 1983 in which radioactive waste liquor was discharged into the sea because the new shift wrongly assumed the contents of a tank to be suitable for discharge into the sea.

Shift handover involves:[24]

1. preparation for handover by outgoing workers;
2. overlap in shift between outgoing and relieving workers to exchange information;
3. review and verification of information by relieving workers after taking over their shift.

In order to avoid any communication breakdown, HSE[25] recommends the following principles where:

1. Higher risk handovers are identified and any associated risks are mitigated.
2. Communication skills for all workforce are developed emphasising the significance of two-way communication during shift handover.
3. Verbal communication takes place alongside written documentation such as checklists for shift handover.
4. Most of the safety critical work is undertaken and completed within one shift where possible.
5. Shift handover is conducted face to face with joint responsibility for the relievers and those being relieved.
6. Tools such as log books and computer displays are provided to facilitate shift handovers.
7. Distinction is made between handover between personnel who are coming on shift after a short break such as 12 hours on/off and those who are coming after a long break such as 2 weeks on/off. The latter must be allowed more time for taking over their shift.
8. An appropriate chain of command is established for shift handover where an individual worker hands over the shift on behalf of a group of workers.

2.13 IOGP Life-Saving Rules

The International Association of Oil & Gas Producers (IOGP) is an organisation established since 1994 with membership from oil and gas producers as well as upstream associations across the world. These members operate in over 80 countries to explore for hydrocarbons at sea and ashore.

IOGP's life-saving rules and other materials are produced to raise awareness of activities that can cause fatalities. These rules were published on 8 April 2013 and focus on affecting workers' behaviours thereby advising individuals to take actions for their and others' safety. These rules are reproduced below from IOGP's publication 'IOGP Life-saving rules' available to download in full from www.ogp.org.uk/pubs/459.pdf. These rules are divided into two sets as (see Table 2.8):

1. Core Rules: a set of eight rules applicable to the oil and gas industry.
2. Supplementary Rules: a set of ten rules that may be applicable to certain parts of the industry's shore based and offshore sectors.

Table 2.8 – IOGP's life-saving rules				
	Rules	Icon	Rules	Icon
Core Rules	1. Obtain authorisation before entering a confined space		5. While driving, do not use your phone and do not exceed speed limits	
	2. Protect yourself against a fall when working at height		6. Follow prescribed journey management plan	
	3. Do not walk under a suspended load		7. Work with a valid work permit when required	
	4. Wear your seat belt		8. Verify isolation before work begins and use specified life protecting equipment	

(continued)

Table 2.8 – (continued)

9. Prevent dropped objects

14. Do not work under or near overhead electric power lines

10. Position yourself in a safe zone in relation to moving and energised equipment

15. No alcohol or drugs while working or driving

Supplementary Rules

11. Obtain authorisation before starting excavation activities

16. Do not smoke outside designated smoking areas

12. Conduct gas tests when required

17. Follow prescribed lift plan

13. Wear a personal floatation device when required

18. Obtain authorisation before overriding or disabling safety critical equipment

Table 2.9 – Categories of IOGP's life-saving rules

	Core Rules	Supplementary Rules
Personal Safety		
Control of Work		
Site Safety		
Driving		

(Source: IOGP Life-saving rules www.ogp.org.uk/pubs/459.pdf)

The above rules are further grouped into four broad categories shown in Table 2.9:

1. Personal Safety
2. Driving
3. Site Safety
4. Control of Work.

Notes

1 Scottish Executive (2004) Guidance on Criminal Penalties in Scotland. [Online] Available at www.scotland.gov.uk/Topics/Justice/criminal/17543/11208 [Accessed 06.06.06].
2 HSE (2014) Slips and Trips and Falls from Height in Great Britain, 2014. Health and Safety Executive [Online] Available at www.hse.gov.uk/statistics/causinj/slips-trips-and-falls.pdf [Accessed 03.02.15].
3 CCOHS (2012) Canadian Centre for Occupational Health and Safety. Fatigue. [Online] Available at www.ccohs.ca/oshanswers/psychosocial/fatigue.html [Accessed 04.02.15].
4 HSE (2008) Guidance for Managing Shiftwork and Fatigue Offshore. Offshore Information Sheet No. 7/2008. [Online] Available at www.hse.gov.uk/offshore/infosheets/is7-2008.htm [Accessed 05.02.15].
5 HSC (1993) ACSNI (Advisory Committee on the Safety of Nuclear Installations) Study Group on Human Factors (3rd report) Organising for Safety. HSE Books, Suffolk, UK.
6 ISF (2006) Implementing a Safety Culture; ISF Safety Culture Leaflet. [Online] Available at www.marisec.org [Accessed 09.07.06] Maritime International Secretariat Services Ltd (Marisec).
7 Heinrich, H. (1959). Industrial Accident Prevention. McGraw-Hill, New York.
8 Cooper, M.D. (1998) Improving Safety Culture: A Practical Guide. J. Wiley & Sons, Chichester, UK.
9 SMART (2006) SMART Goals. [Online] Available at www.projectsmart.co.uk [Accessed 14.06.08].
10 Step Change in Safety (2003) Leading Performance Indicators: Guidance for Use. Step Change in Safety, UK.
11 Minerals Council of Australia (2003) Positive Performance Measures – A Practical Guide. Minerals Council of Australia, Sydney.
12 Duncan, W.J. (1978) Organisational Behaviour. Houghton Mifflin Company, Boston.
13 WMA/OPITO (2012) MIST course notes. Southampton, UK.
14 Step Change in Safety (2005) Personal Responsibility for Safety Guidance. [Online] Available at www.stepchangeinsafety.net [Accessed 07.11.14].
15 MCA (1999) Code of Safe Working Practices for Merchant Seaman. MCA, Southampton, UK.
16 HSE (2005) Guidance on Permit-to-Work Systems, HSG 250. [Online] Available at www.hseni.gov.uk/ [Accessed 31.01.15].
17 Khalique, A. (2010) Shipboard Safety Officer Notes. Witherby Publishing Group Ltd, Scotland.

18 Lawrence, D.P. et al. (2004) Risk Assessment. [Video] Maritime Research Ltd and Templar Film and Television, Devon.

19 Glossop, M. et al. (2000) Review of Hazard Identification Techniques: Project Report by Health and Safety Laboratory. [Online] Available at www.hse-databases.co.uk/research/hsl_pdf/2005/hsl0558.pdf [Accessed 31.07.08].

20 O'Conor, D. (2002) Managing Health and Safety. Scitech Educational, Broadstairs, Kent.

21 OPSI (2006) SI 1999 No. 3242 – The Management of Health and Safety at Work Regulations 1999 – Schedule 1. [Online] Available at www.opsi.gov.uk/si/si1999/19993242.htm#sch1 [Accessed 06.08.06].

22 MCA (1998) MGN 20 (M + F) – Implementation of EC Directive 89/391 – Merchant Shipping and Fishing Vessels (Health and Safety at Work) Regulations 1997. MCA, Southampton.

23 ABS (2013) Management of Change for the Marine and Offshore Industries. American Bureau of Shipping. [Online] Available at www.eagle.org [Accessed 07.02.15].

24 Lardner, R. (1996) Effective Shift Handover. Health and Safety Executive. Offshore Safety Division. [Online] Available at www.hse.gov.uk/research/otopdf/1996/oto96003.pdf [Accessed 07.02.15].

25 HSE (2015) Human Factors: Shift Handover. [Online] Available at www.hse.gov.uk/humanfactors/topics/shift-handover.htm [Accessed 07.02.15].

3

Travelling Offshore

Whilst the offshore industry may be considered as having one of the most hazardous workplaces due to the location of the offshore installation, travelling to this workplace has its own complications. There are two modes of transportation available to and from the offshore installations i.e. travel by boat or by helicopter. Both these modes of travel have their own pros and cons which are looked at in this chapter.

3.1 Travelling by Helicopter

Helicopters are considered to be the workhorse of the offshore oil and gas industry. Generally, the offshore workers spend 2–4 weeks on the installation after which they are onshore, either on leave or on another job. After spending their time on-board the offshore installation, the workers' preference is to reach home as quickly as possible; hence the helicopters provide a very quick means of transportation. In addition to getting them home quickly, the helicopters also contribute to considerable savings in terms of employees' reduced costs for time at work. Other means of transportation i.e. by boat have significant limitations

Figure 3.1 Helicopter on helideck of an offshore platform (Source: BP plc)

in terms of the speed at which they can move; therefore they do not provide the same time-saving benefits.

The helicopters used for offshore passenger transport in the UK are usually 18 seats but some smaller aircraft may also be used. The passengers are required to have undertaken BOSIET (Basic Offshore Safety Induction and Emergency Training) or FOET (Further Offshore Emergency Training) approved by OPITO. These courses are mandated by the offshore operators to ensure the safety of travellers. Occasionally, some operators may allow temporary workers or contractors to undergo only HUET (Helicopter Underwater Escape Training) prior to travelling offshore by helicopter. Details of this training are given in Chapter 11. In addition to undertaking these courses, passengers must follow other strict rules such as:

- Wear helicopter transit suits and lifejackets fitted with built-in or additional EBS (Emergency Breathing System), etc.
- Passengers whose shoulder width is 22 inches or more are considered 'extra broad' and therefore are required to sit next to a large helicopter window that will provide them with a means of escape in case of emergency.

Over 2 million passengers are carried offshore annually by helicopters in the UK offshore sector alone. Due to the large number of flights and number of personnel involved and the hostile environment in which these aircraft operate, the CAA (Civil Aviation Authority) regulates the helicopter industry quite strictly. Over the years, due to accidents with the helicopters and advancement in technology, new types of helicopters have been introduced in the industry. A few examples of helicopter types currently utilised for passenger transport in the industry are given in Table 3.1.

Table 3.1 – Common helicopters used in the offshore oil and gas industry[1]		
Type	Seating Capacity (excluding pilots)	Range (nautical miles)
AS332 (Super Puma)	19	454
Air Bus AS365 (Dauphin)	11	427
Eurocopter EC155/EC225	12/19	427/448
Sikorsky S61/S76/S92	18/14/19	450/449/750
AgustaWestland AW139	15	573

Depending upon the location of offshore installation, a nearby airport can provide heliport facilities. Examples of heliports in the UK are as follows:

1. Aberdeen Heliports has three terminals:
   - Bond Heliport Terminal;
   - Bristow Heliport Terminal;
   - CHC Heliport Terminal (Scotia).
2. Scatsta Heliport (Shetland Island);
3. Blackpool Heliport (Squires Gate);
4. Humberside Heliport;
5. North Denes Heliport (Great Yarmouth);
6. Norwich Heliport.

3.1.1 Check-In Procedures

Most heliports operate in a similar fashion to the commercial airports. However, due to the nature of the flight, certain additional restrictions have to be placed on passengers to ensure the safety of those on-board and the aircraft.

Generally, the flight schedule will be provided to the passengers but the latest information can always be checked by contacting the heliport or the employers' flight/logistics coordinator. After ensuring the correct flight time, passengers should arrive prepared for travel and 'in time' to go through check-in procedures. Preparations must include:

- Travel documents such as passport or other equivalent document. Their remaining validity requirements vary and must be checked with the employer/duty holder prior to travel.
- Authorisation from the duty holder, OIM or employer to confirm details of the person travelling offshore. Any special visitors e.g. dignitaries, etc. will also require evidence of authorisation or a special pass. Certain offshore operators issue passes to their employers or contractors hired by them. The following types of passes are in use:
  - travel safety by boat (for personnel travelling by boat);
  - full offshore pass (for full time offshore employees);
  - restricted offshore pass (for boat or helicopter crews);
  - infrequent traveller pass (for personnel who do not travel on regular shift patterns);
  - special pass (for dignitaries)
- A valid offshore medical certificate. Personnel with any disability or disease will not be allowed to travel offshore.
- Personnel under the influence of drugs or alcohol will not be allowed to board the helicopter.
- Basic Offshore Safety Induction and Emergency Training (BOSIET) or Further Offshore Emergency Training (FOET) certificate in the UK. Overseas equivalent certification may be accepted but this must be checked prior to travelling.
- Some offshore operators also require OPITO approved Minimum Industry Safety Training (MIST) certificate or another equivalent certificate.

- Vantage Card (if the employer/operator is a member of Vantage POB service).
- Passengers must arrive on the heliport in loose fit casual clothing preferably with lace-up shoes. Sandals or slippers are not allowed.
- Personal medicines can be carried but must be declared to the heliport security. On arrival at the installation, these must be reported to the medic or the OIM.
- PPE including safety shoes, goggles, helmets, overalls and hearing protection must be carried.
- Helicopters operate with strict baggage limitations. Maximum weight allowed in most cases is 10–15 kg including any personnel possessions such as laptop, eBook reader or other electronic equipment. Baggage should be carried in a soft sided bag (plastic bags are not allowed). Laptops should be packed in a separate protective case to avoid damage. Excess baggage is not permitted at all due to limited space travelling in any one flight.
- Some workers may need to carry equipment or tools with them. This must be arranged beforehand and proper authorisation obtained. The following items are prohibited for carriage on flights:
  - weapons or explosives of any type including pyrotechnics;
  - flammable liquids such as lighter fuels;
  - pressurised gas containers including aerosols;
  - corrosive, radioactive or toxic materials;
  - matches and cigarette lighters;
  - fishing equipment;
  - knives or other sharp tools unless prior authorisation is obtained.

Figure 3.2 Offshore workers boarding a helicopter

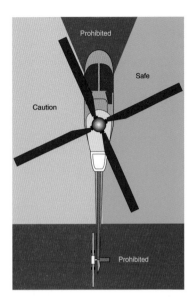

Figure 3.3 Helicopter safety zones

- Mobile phones or other transmitting devices, cameras, radios, i-Pods, MP3 players, etc. may not be allowed on-board offshore installations; hence passengers must check the requirements before bringing them to the heliport. In some cases, these may be allowed but passengers are not allowed to use them during the flight.
- Loose paper such as magazines or newspapers is not allowed due to the possibility of it being ingested by the engines.
- A survival/transit suit will be supplied at most of the heliports. Passengers may be asked to don this suit after obtaining the boarding pass and going through customs/immigration. Immediately prior to departure, a safety video is also shown to all passengers at most of the heliports.
- When boarding the helicopter, it should always be boarded via the 'safety zone' regardless of whether the rotors are turning or not (Figure 3.3).

3.1.2 Arrival on Installation by Helicopter

All passengers must remain seated with seat belts/harnesses fastened until advised by the pilot or Helideck Landing Officer (HLO). Any instructions given by personnel coordinating disembarkation must be strictly followed. Helideck crew will unload baggage from the helicopter. Upon instructions from the HLO, passengers can collect their baggage and proceed to the reception or administration desk. On unmanned installations, passengers will be required to unload baggage by themselves.

Regular workers, commonly known as residents, will be informed about their cabin allocation and given information about muster stations. Short-term visitors, referred to as 'day visitors', will also be informed about their muster stations, emergency procedures and the contact details of personnel relevant to their job. For first-time visitors or others returning after a gap of 4–6 weeks, a health and safety induction will be organised. Upon completion of these

Figure 3.4 A helicopter landed on the helideck of an offshore installation

tasks, personnel carrying prescribed medicine must see the medic to establish arrangements for their medicines.

The procedure for departure is similar to the check-in procedure except that the reception or administration office on the installation will issue the boarding passes.

3.2 Travelling by Boat

Travelling to an offshore oil and gas installation by boat follows procedures similar to some extent to travel by helicopter. The passengers are required to arrive earlier than the departure time, e.g. 30 to 60 minutes in most of the terminals.

Passengers are required to have an offshore travel pass as described in the travel by helicopter section. Baggage is checked by security personnel prior to departure with the same weight limitations as for helicopter travel. After checking in, security and immigration clearance, passengers will be given a safety briefing without which boarding is not allowed. Prior to boarding the boat, passengers are required to don their safety helmet, safety shoes and lifejacket.

Passengers may be required to undertake the OPITO approved 'Travel Safely by Boat' training prior to being allowed to travel by boat. The course requirements are discussed in Chapter 11 in detail.

3.2.1 Arrival on Installation by Boat

On manned offshore installations, similar to the HLO, a Boat Landing Officer (BLO) will be designated to oversee the transfer of personnel to and from the boat. Travelling by boat is quite common in tropical areas and in other areas where weather conditions permit, especially the height of swell. Various methods of transfer from boat to the installation are used in the industry. These are discussed in the next subsections.

3.2.2 Transfer by Rope

If the boat or the installation is not fitted with a mechanical means of transferring personnel to or from the boat, a common method used is known as 'swing rope transfer'. In this method, a strong 'knotted' rope attached to a high point on the installation is used for transfer. The transferee holds the rope with both hands and swings from vessel to the installation or vice versa. In order to ensure safety, standard operating procedures must be complied with at all times. For example,

- The boat landing area may only allow access to a limited number of personnel at any one time.

Figure 3.5 Workers waiting for transfer by swing rope

- Two crew members must be present on each side i.e. the installation and the boat to assist the passengers.
- Crew must check all hardware and surrounding area prior to use.
- Baggage should be transferred before passengers.
- Transferees must securely don their PPE e.g. lifejacket, safety boots and hard hat. Pockets should be empty or secured and personnel should not have any loose items on them.
- Patience and timing is the key to success of this operation, hence all personnel must proceed with due diligence.
- Transferees must have both their hands free, holding the rope at a knot just above their eye level, ensuring clearance above any obstructions in their path and keeping in mind the movement of the boat due to waves.
- Keep the rope clear of the legs otherwise it may cause problems during landing.
- Means of communication, particularly signals to 'abort' the operation, must be agreed and conveyed to all personnel involved.
- If the transferees are not confident about undertaking transfer by this method, they should never be forced.

3.2.3 Transfer by Basket

Personnel can be transferred between an offshore installations and vessels using a specially designed personnel transfer basket. Due to the critical nature of this operation, a decision must be made jointly by the master of the vessel with which the transfer is taking place and the OIM.

Generally this method of transfer is utilised where no other means are available or impractical due to safety reasons. The risks of injury in this method are considerably high and include falling from the basket, impact on landing, trip hazards when accessing the basket, etc. When undertaking these transfers, the following precautions should be taken into consideration:

- Standard operating procedures must be followed.
- Approval from OIM and vessel master should be obtained prior to commencing operations.
- Competent crew should be available both on installation and the vessel.
- All equipment and hardware should be checked prior to use.
- The ability of the vessel to maintain position should be considered in relation to the prevailing weather conditions.
- All personnel being transferred must be attired appropriately including PPE such as hard hat, safety shoes and lifejacket and if available a Personal Locator Beacon (PLB).

Figure 3.6 Transfer to offshore platform by basket

Figure 3.7 FROG personnel transfer capsule

- The basket and crane including all accessories being used must comply with the legal requirements for their use such as with the Lifting Operations and Lifting Equipment Regulations (LOLER) 1998. All items need to be certified to 'man-riding' equipment requirements.
- A clear route for the movement of the basket must be established and followed.
- Competency of the crane operator and means of communication between the signaller (referred to as the banksman in the offshore industry) and the crane driver must be established and tested prior to commencing the transfer.
- If the transferees are not confident undertaking transfer by this method, they should never be forced.

A number of crane transfer devices are used in the industry. For example,

- special personnel carrier such as FROG/TORO personnel transfer capsule developed by Reflex Marine (Figure 3.7);
- Billy Pugh Personnel Transfer net;
- SAFETRANSFER basket formally known as ESVAGT's personnel basket.

Whilst these devices are safer than some of the more basic ones, the generic safety considerations stated earlier will continue to apply for these devices as well.

3.2.4 Transfer by Gangway or Bridge

In many cases, a gangway or a bridge may need to be utilised as the primary means for the transfer of personnel between the boat and the installation. This mode of transfer does not include use of fully enclosed gangways or bridges.

Figure 3.8 Gangway between accommodation unit and production platform

In addition to the general precautions, personnel must wear appropriate PPE i.e. hard hat, safety shoes and lifejacket. General precautions for gangway transfer are below:

- Transfer must be supervised by personnel familiar with standard operating and emergency procedures.
- Inspection must be carried out prior to transfer.
- Capacity should never be exceeded.
- Baggage should not be carried on the gangway. Alternative means such as a baggage container should be used.
- Safety nets should be rigged where a risk of falling overboard from the gangway exists.
- Access should be free from obstructions.
- Sufficient illumination should be provided at all times.
- Any alarm systems fitted to monitor the angle of inclination (30 degrees is the upper limit) or movement of the bridge/gangway should be tested regularly.

3.3 Medical Care Offshore

In the UK, the Health and Safety at Work etc. Act 1974 requires employers to ensure, so far as is reasonably practicable, the health and safety of workers and others. The Management of H&S at Work Regulations 1999 supplement this requirement by requiring employers to undertake risk assessments to identify hazards and control risks to workers and others. In order to achieve the latter, employers are required to appoint H&S officer(s) supported by offshore medics, first aiders and others. Furthermore, the Offshore Installations (Safety Case)

Regulations 2005 require these arrangements to be included in the Safety Management System (SMS), which is a part of the safety case for each installation. Further requirements are specified in Offshore Installations and Pipeline Works (First Aid) Regulations 1989. Under these regulations, two important roles need to be considered:

1. An offshore medic is a person who holds a current Offshore Medic Certificate issued by an organisation approved by HSE. An installation with 25 or more persons at any one time is required to have ONE offshore medic. Under normal circumstances, the offshore medic is authorised to deal with routine minor ailments such as colds and coughs and administer medication.
2. An offshore first aider is a person who holds a current offshore First Aid Certificate issued by an organisation approved by HSE. An adequate number of first aiders is required at all times for a regularly manned installations. This number is determined by the type of work being undertaken and associated risks as well as the number of personnel employed on the installation. On installations without an offshore medic due to number of employees being less than 25, at least two qualified first aiders should be provided.

All normally attended offshore installations, pipe laying barges and heavy lift vessels used in offshore construction, repair, dismantling or other related activities are required to be provided with a sick bay. The sick bay facilities, medical supplies and equipment should be suitable to accommodate the number of personnel normally present on the installation. Whenever personnel are present or working on installations which are not normally attended, they must have access to a resuscitator with an adequate oxygen cylinder supply and sufficient quantity of first aid materials based on the safety case assessment.

3.4 Environmental Awareness and Waste Disposal

Most offshore operators and therefore the installations are ISO 14001 compliant to reflect an operator's management of environmental risks and preserve the environment whilst meeting its business objectives. An organisation certified under ISO 14001 standards is required to demonstrate that it:

• is fully aware of the impact of its activities on the environment;
• develops policies that reduce harmful impacts on the environment;

- accepts responsibility through implementation of an Environmental Management System (EMS);
- involves employees in implementing the requirements of the EMS through raising environmental awareness and development of competencies through training.

In order to achieve the above, the operator benefits from reduction of waste, increased efficiency/performance, compliance with legislation and therefore increased profits and employee confidence. Each operator will cover certain specific areas to gain the foregoing benefits. These areas include:

1. materials handling including chemicals;
2. marine management;
3. hazardous or non-hazardous waste;
4. air emissions;
5. hydrocarbons;
6. energy;
7. water.

The Environmental Protection Act 1990 covers the protection of the environment in the UK. Furthermore, the Offshore Petroleum Activities (Oil Pollution Prevention and Control) OPPC Regulations 2005 deal with discharges of oil resulting from offshore oil and gas exploration and production activities. The UK is also a signatory to the Convention for the Protection of the Marine Environment of the North-East Atlantic (the OSPAR Convention) 1992. According to this convention, dumping (i.e. any deliberate disposal in the maritime area of wastes or other matter from offshore installations) is forbidden. Collectively, this legislation provides necessary requirements to prevent pollution and also gives powers to relevant bodies to inspect, investigate and take action for any contravention of the requirements. These requirements also define what, when and how to report any accidental or unplanned discharges of crude oil, diesel oil, lubricating oil and hydraulic oil or chemicals that may cause pollution. In addition to the above, all employees must strictly follow the procedures laid down in the EMS for environment protection and pollution prevention.

3.4.1 Waste Management and Prevention of Pollution

Like any other industry, the offshore industry also generates different types of waste during exploration and production activities. Additional domestic waste is generated whilst workers live and work on-board. Waste materials generated on-board include paper, glass, cardboard, metals, plastic, oil, sludge and chemicals.

Offshore waste has to be managed as required by legislation. Examples of applicable legislation are given in Table 3.2.

Table 3.2 – Offshore pollution prevention examples

Type of Waste	Applicable Legislation	Summary of Requirements
Garbage	International Convention for the Prevention of Pollution from Ships (MARPOL) 1973/78	MARPOL Annex V deals with different types of garbage generated on-board ships or offshore installations. It also specifies the requirements for disposal of various types of garbage. This annex prohibits disposal of any garbage into the sea except comminuted food at a distance of over 12 nautical miles from shore. All records related to garbage are required to be maintained in a Garbage Record Book and disposal must be carried out according to the installation specific Garbage Management Plan (GMP). A number of different types of placards are required to be displayed around the installation to give workers knowledge of GMP requirements.
Hazardous Waste	• Special Waste Regulations (Scotland) • Hazardous Waste Regulations (England and Wales) • Environment Protection Act (EPA) 1990	Hazardous waste must be transferred ashore for disposal. This transfer requires a 'transfer note' for each disposal instance. Description of type and quantity of waste needs to be recorded to ensure compliance with the requirements.
Oily Wastes	IMO Resolution A.863(20) – Code of Practice for the Carriage of Cargoes and Persons by Offshore Supply Vessels	• Various operations on-board generate mixed/contaminated fluids that must be discharged onshore. These wastes need to be brought ashore for proper disposal. • Oil & Gas UK established the Oil Spill Prevention and Response Advisory Group (OSPRAG) to deal with ongoing changes in the industry.
NORM	Radioactive Substances Act 1993 (RSA)	Naturally Occurring Radioactive Materials (NORM) may contaminate the water produced during drilling or production stages. This water may contain NORM such as barium, uranium, radium or thorium. If discharged ashore, radium, for example, may bioaccumulate in marine organisms and then pass to humans. Fortunately, very low concentrations of radium are passed to marine organisms as a result of offshore oil and gas activities. As a result, NORM waste is currently exempted from the disposal requirements of RSA. However, this may change in the near future if facts contrary to the foregoing evidence become available.
Atmospheric Emissions	• MARPOL 73/78 Annex VI • Petroleum Act 1998 • Environmental Protection (Controls on Ozone Depleting Substances) Regulations 2010	Emissions from offshore installations such as CO_2, CO, NOx, SOx, N_2O and ozone depleting substances must be controlled as per the requirements given in the MARPOL convention or other applicable legislation.

3.5 Hygiene

Personal hygiene is the process of cleansing one's own body to ensure good health and wellbeing and protect from infections or other diseases. A small effort on a regular basis can make a significant difference on an individual as well as organisational basis.

In any case, high standards of personal hygiene will help increase self-esteem and confidence. Poor hygiene will on the other hand not only have personal health implications but also have social and psychological aspects.

Keeping in mind that the offshore installations are a work as well as a living place for workers, the significance of personal as well as workplace hygiene becomes even more important. When ensuring hygiene in an offshore work environment, the following factors will act as guidance:

- Work, food storage and cooking areas and living quarters should always be kept clean.
- A good standard of ventilation must be maintained at all times.
- Personnel should take at least one bath or shower every day.
- Appropriate gloves and other PPE should be used to prevent contact with any contaminant chemicals.
- Hand and eye wash facilities should be used as recommended.
- Work clothes and shoes should not be worn in living quarters.
- Contaminated clothing and PPE should be treated according to the recommended procedures, and disposed off if required.
- Smoking restrictions should be strictly complied with.
- EMS requirements including waste disposal should be complied with to improve hygienic conditions and prevent pollution.
- Food hygiene should be given due significance as it may easily be the most common means of spreading microbiological organisms.
- Any symptoms such as a skin rash indicating infection or other disease must be reported to the supervisor immediately.

Note

1 Sikorsky S-92 Helicopter Brochure. Available at www.cougar.ca/media/images/brochures/cougar-s92.pdf [Accessed 16.10.14].

4

Manual Handling and Working at Height

About a quarter of the injuries in the offshore oil and gas industry are caused during moving or lifting objects. These injuries can easily be avoided through simple and cost-effective preventive measures. Improper lifting, failing to use appropriate equipment and unsafe work practices are the most common reasons for injuries such as strains, sprains, fractures or contusions (bruises).

In the offshore industry, the most common workplace where manual handling related injuries are likely to occur include the rig floor operations where handling heavy tubulars and other associated equipment require extreme care alongside a high level of physical fitness. A minor carelessness in handling equipment can lead to an injury.

4.1 Mechanics of Manual Handling Injury

Most manual handling injuries are related to damage to the spine (also known as spinal column, vertebral column or the backbone). The 'S' shaped structure of the spine (Figure 4.1) provides the ability to walk, run or even sleep. Arms, legs, chest and head are all attached to the spine. This spine shape allows it to absorb the shocks resulting from normal activities such as walking or running. Any damage to the spine will affect any movement the body parts can make. It is therefore important to understand the basics of this structure and how easily it can be damaged to prevent its exposure to a situation that could cause damage.

The spine consists of several bones known as 'vertebrae'. Between any two vertebrae, rubber-like discs called 'facets' are placed to offer flexibility to the spine. These discs are made of two parts, an outer skin made of strong fibres and an inner part made of jelly-like material. When the outer part of this disc ruptures, it is known as 'herniation', a common manual handling related injury (Figure 4.2).

Back pain, referred to as 'dorsalgia' in medical terms, is the pain felt in the back originating from nerves, muscles, bones or other structures in the spine. The spinal cord runs through a passage within the spine. As a result of slips and

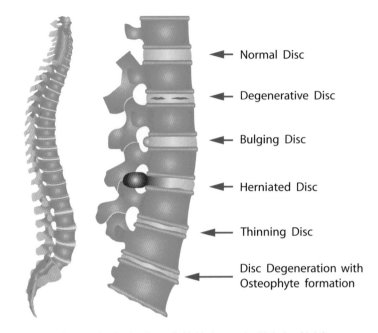

Figure 4.1 A normal spinal column (left); Various spinal injuries (right)

Spinal disc herniation

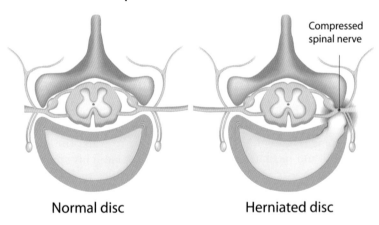

Figure 4.2 Manual handling injury

falls, the spinal cord may be damaged or internal bleeding over a number of days in or around the spinal cord may be the cause of pain. Generally, there are two types of low back pain known as lumbago and sciatica. Lumbago is the pain caused in the muscles and joints of the lower back whereas sciatica is damage to the nerve that runs from the back of the body to the feet which may be because of a herniated or slipped disc.

4.2 Management of Manual Handling Operations

In the UK, the Manual Handling Operations Regulations (MHOR) 1992 provide detailed requirements for risk assessments for manual handling tasks. The regulations apply to all activities involving transporting or supporting a load that includes bodily force for lifting, putting down, pushing, pulling, carrying or moving the load. Employers are required to avoid manual handling if possible, and assess and control the risks if manual handling can't be avoided. Employees are required to follow the agreed procedures, ensure supplied equipment is used fully and properly, and inform supervisors of any new hazards. Further guidance includes the following:

1. The employer must have a policy covering manual handling operations.
2. If any operation involves risk of injury, avoid manual handling as far as is reasonably practicable.
3. Consider the use of handling aids such as pallet trucks or mechanically powered lifts, etc. keeping in mind that these aids must reduce the existing risks and avoid creating new ones.
4. Assess and control ALL identified risks noting that hazard identification is the most important process for controlling the associated risks. Eliminating risks may not always be possible as the rule is to 'control risks so far as is reasonably practicable'. Once appropriate assessment has been carried out, records must be kept for all significant findings.

Figure 4.3 Manual handling techniques

5. Other factors to consider include:
   a. Ergonomics of the items involved in lifting operations such as the shape, size, weight and lifting points fitted on the item to be lifted, the area through which it is being moved, the design and limitations of any mechanical lifting device being used.
   b. Inappropriate position of the body may lead to injuries such as sprains or strains. Hence the possibility of twisting and bending of the body during the whole operation must be considered.
   c. Training for lifting operations including an awareness of the causes of manual handling related injuries, correct techniques for lifting operations. This includes training for use of any lifting aids being used.
   d. Communication during all stages of the operations.

Whether one person is lifting a box, multiple persons are lifting a heavier load or a lifting aid is being utilised to carry larger and heavier items, the basic techniques to ensure safety of personnel remain the same. This is based on the following steps described as PAGEUP:

1. **Plan** the lift, ensuring awareness of risks and their controls, use of aids or help from additional persons, etc. Plan from lifting the load to landing it including the route through which it will be carried, any obstacles in the way such as stairs or turns.
2. **Adopt** a proper and stable position to lift the load by positioning feet on the sides of the load to provide a stable stand for lifting (Figure 4.3).
3. **Grip** the load securely and firmly keeping it close to the body. It will feel lighter in comparison with lifting it with stretched arms as some support can be obtained from keeping it close to the body. Ensure suitable lifting points, if none available then use a

Figure 4.4 Management of manual handling

strong grip to hold the weight which will cause less tiredness to fingers than a loose grip. If the position of hand/fingers needs to be changed, then find a suitable resting point.

4. **Execute** the lift ensuring continuity of the precautions mentioned above. These must be adhered to throughout the process and not just when lifting the load.

5. **Use** mechanical aids or additional help. If the load is beyond one person's capacity or too strenuous, then don't try lifting it without additional help. Remember, risks must be controlled, if they can't be then the task must be postponed until they can be controlled. One bad attempt could render a person incapacitated for a prolonged duration, even for the rest of their life.

6. **Posture** helps transfer the load from the arms to the body without causing it to concentrate at any specific location of the body e.g. shoulder, knees or elbows. A crouched position with bent knees and back remaining in a natural curved position is considered best to begin the lift. Fully flexed back, hips and knees are considered less helpful in manual handling. When putting the load down, the lifting procedure is reversed. Risks such as trapped fingers or putting the load down on an unstable surface may still lead to injury if the whole operation is not fully planned.

4.3 Manual Handling Risk Assessment

It will be almost impossible to produce a standard set of rules for all manual handling tasks since the physical capacity to lift any item varies with each individual. However, if the basic concept of 'ergonomics of work' is applied to all activities, many of the underlying issues can be resolved. Ergonomics can be described as 'fitting the task to the person instead of fitting the person to every task' meaning that each individual has a different set of capabilities (or limitations) that compromise their compatibility with each task. It is therefore important that a risk assessment is carried out to provide a balance between the worker and the task they are being assigned. In all situations where manual handling cannot be avoided, the employer has a legal responsibility to ensure risk assessments are carried out. Furthermore, even if the lowest possible capability for any individual amongst the workforce is established, no two tasks will be the same due to variable factors such as the time of the day, level of fatigue in an individual, type of item being lifted and so on.

Generally, work activities involving manual handling can be categorised on the basis of difficulty observed during lifting operations to establish general requirements that apply in all circumstances. This should then be incorporated into each work activity at the design stage. When the task is then being carried out, each manual handling operation should be looked at on a case-by-case basis and any new or additional risks identified and controlled. The process of risk assessment is discussed in detail in Chapter 2 but here the hazards that need to be considered for manual handling activities are divided into four categories:

1. Plant – the load that is being handled manually. Examples of hazards to look for in this category are:
   a. position of the object when being lifted, carried or landed;
   b. posture change requirements such as twisting the body, pushing/pulling, grip vs movement of the item, etc.;
   c. weight and size of the object;
   d. any lifting points fitted;
   e. contents of the objects e.g. chemicals, sharp edges or hot items.
2. Place – the workplace where the manual handling task is being carried out to include the examples below:
   a. reaching upwards or downwards to lift the object;
   b. obstacles such as uneven deck/floor, stairs or ramps;
   c. distance through which it has to be carried;
   d. available space through which to manoeuvre;
   e. external impacts on the operation such as rolling, pitching of the vessel/platform, effect of weather and illumination of the area through which the object is being carried, etc.;
   f. restrictions due to PPE in use.
3. Competency of the individual in relation to the task. This should include the physical ability of the person to perform the task. For example, a 65-year-old person may not have the same physical ability as a 23-year-old. A person who has been fully trained and is an experienced 'manual handler' will be considered more competent than a person who is new to the job.
4. System – some of the questions that need to be asked to ensure availability and implementation of a safe system include:
   a. Has the system considered the three foregoing categories of hazards i.e. plant, place and competency?
   b. Has a risk assessment been carried out?
   c. Are the workers informed about the outcomes of risk assessment?
   d. Are all workers involved in the task given full brief (toolbox talk)?
   e. Is appropriate PPE available and being used?

4.4 Basic Manual Handling Technique

Figure 4.5[1] shows a basic guideline to identify when a 'pre-lift' self-assessment will be required in manual handling lifting and lowering operations. Position of the load given in the boxes with respect to the body of the handler gives an indication of the weight that can be lifted safely. If the movement of the load during a lift falls in more than one box, then the lowest figure in the box should apply. In all cases of doubt, a detailed risk assessment must be carried out to ensure safety.

For example a male worker intends to lift a 30 kg weight placed on the floor. This weight can be kept close to his body and needs to be lifted to the knuckle height. In this case, his hands will go through more than one box but the smallest guide weight box 10 kg will be used to limit the weight that can be lifted from floor to knuckle height. Since the weight is more than the guide weight, a detailed risk assessment is necessary to ensure sufficient control measures or else the risk of injury remains quite high.

The guideline weights assume:

- this manual handling task takes place during normal conditions;
- the object being lifted is of normal dimensions;
- the object provides reasonable grip;
- the person lifting the weight is not carrying out this task repetitively for more than 30 times per hour;
- the person maintains a steady posture throughout the lift.

Any variations from the above will require a detailed risk assessment. As a general rule, if the handler is required to twist their body, the guide weights should be reduced by 10 per cent for a 45 degree twist and a further 20 per cent for a 90 degree twist. If the handler repeats the task:

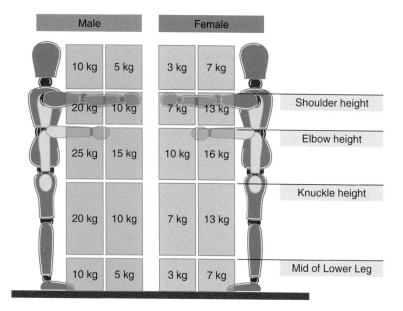

Figure 4.5 Manual handling weight limits (Source: HSE Publication INDG 143)

- twice every minute, reduce weight by 30 per cent;
- 5–8 times per minute, reduce the weight by 50 per cent;
- more than 12 times per minute, reduce the weight by 80 per cent.

4.5 Lifting and Handling Aids

As seen in the discussion about manual handling in the foregoing section, if the risk assessment indicates a requirement to control risks, the use of lifting and handling aids could be one of the control measures. These aids not only reduce risks but also save considerable time in completion of a task. Simple aids such as sack trucks, trolleys and wheelbarrows can be easily acquired and used. In the selection of lifting and handling aids, the following factors should be considered by the employer:

- Employees should be involved during risk assessment and selecting control measures keeping in mind that the perception of hazards and associated risk may change on the basis of workforce competencies.
- Further advice on the suitability of lifting and handling aids can be obtained from the manufacturer of the aid. This advice can include information like safe working load, the number of personnel required to deploy the aid and any specific training required in its use, etc.
- Ensure that the aid is approved for use, normally indicated by the CE mark.
- Handling aids also require inspection and maintenance without which they should not be used. Most of these aids may come under LOLER or PUWER regulations. Examples of equipment covered under these regulations include chains, ropes, slings, shackles, harnesses, eyebolts, fork lift trucks, jacks, mobile elevating platforms, axle stands, vehicle tail lifts and passenger lifts, etc.
- Each handling aid will have certain restrictions and limitations which must be considered when deploying it to avoid any new risks.
- Suitability of each handling aid should be checked for the task at hand.
- Where a risk assessment is required for a manual handling task, one of the control measures should be supervision by a suitably qualified and competent person, especially if the handlers are new to the task.

Some examples of lifting aids are given in Figures 4.6 and 4.7.

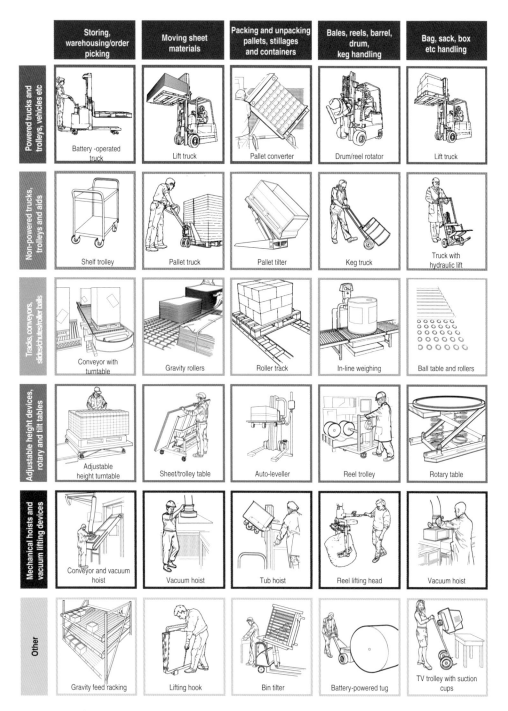

Figure 4.6 Lifting and handling aids (Source: HSE.gov.uk)

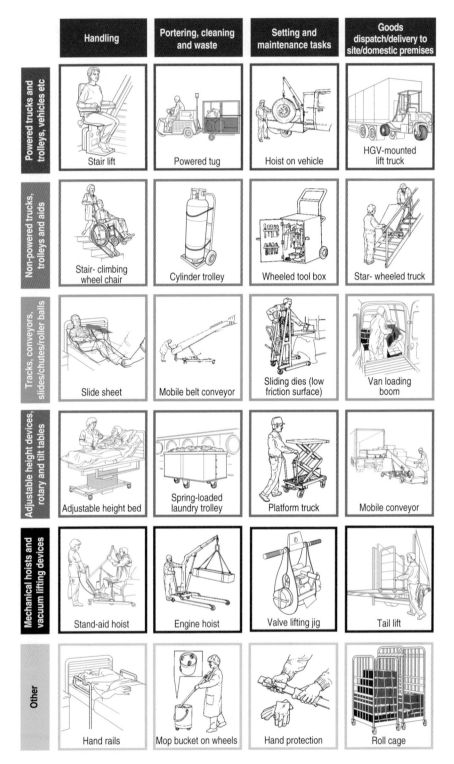

Figure 4.7 Lifting and handling aids (Source: HSE.gov.uk)

4.6 Manual Handling in Teams

There may be many situations where a one person lift is not possible due to the nature of the task. In such situations, more than one person may need to be involved for the safety of operations. The basis weight guide given earlier in this chapter does not change by the addition of a person. When additional personnel are added to the team, additional weight equivalent to two-thirds of that given in the weight guide can be added but this should not be treated as a mandatory requirement. A safely completed lift means a second lift could be undertaken. Compromising safety may compromise the ability of the team to undertake any further tasks, hence it is better to be safe than sorry.

Ideally, the handling team should nominate a leader who will coordinate the task. If, for example, an object of dimensions 1 m x 2 m x 1.5 m weighting 45 kg was to be lifted, then a four person lift could be deployed. These four people should ideally be of the same height to keep the lift at a level height from the deck. One person can be nominated as the leader. A quick 'toolbox' talk at the beginning will clear any ambiguities such as for lifting, moving, turning, stopping and landing the object. Communication about all of these factors should be carried out loudly by the leader. For example, we will lift at the count of 3. Rushing through the task may compromise

Figure 4.8 In a group lift, always nominate a leader

Figure 4.9 Working at height
(Source: BP plc)

Figure 4.10 WAH is common for
scaffolders

safety, hence two-way communication between the team members should be maintained throughout the lift. If any team member feels uncomfortable due to a change in circumstances, they should raise it immediately to other team members. A contingency plan for any such eventualities should be agreed in advance.

4.7 Working at Height

Falls from heights continue to cause thousands of injuries and dozens of fatalities in the offshore oil and gas industry.

Working at Height (WAH) means any workplace without any protection from which a person could fall a distance with potential to cause an injury. The following situations fall under the WAH:

- anyone not standing on level ground or at the deck level;
- working inside a tank;
- working near an opening such as a tank opening, an open hatch or a stairway.

PUWER and LOLER regulations now require that any work at height should only be undertaken if it cannot be avoided. Safety of workers should always be the primary concern and basic considerations for hazard identification, risk assessment and control apply in all circumstances where any task falls under work at height.

The WAH activity must be properly planned, supervised and carried out by suitably qualified and trained i.e. competent personnel using appropriate PPE under the Work at Height Regulations 2005 (WAHR). These regulations cover the following areas related to WAH:

- existing places of work and means of access for work at height;
- collective fall prevention (e.g. guard rails and toe boards);
- working platforms;
- collective fall arrest (e.g. nets, airbags etc.);
- personal fall protection (e.g. work restraints, work positioning, fall arrest and rope access);
- ladders and stepladders.

The risk assessors will note that when applying the hierarchy of controls, addition of some controls may mean that some other controls are not required. For example, if scaffolding is erected in such a way that the work is no longer considered to be hazardous and the risk of fall does not exist, then personal fall protection measures may not be required.

Appropriate equipment includes the Mobile Elevating Work Platform (MEWP) where specific training is required for the operator. Special fall protection equipment is available and must be used where required. Employees are required to:

- report any hazards they identify;
- following any training, use the equipment supplied by the employer in the most appropriate way.

Both employers and employees are required to ensure safety is never compromised. The planning should always include contingencies for any emergency situations including procedures for rescue for those stranded during WAH tasks. Whilst carrying out risk assessment, due regard should be given to prevalent weather conditions which may have a significant impact upon the control measures put in place for the task at hand. It is vital to note that the workplace for WAH activity must also include areas surrounding the main work area such as means of access. In cases where a fall cannot be prevented, the regulations require the distance of the fall to be reduced so far as is practicable in addition to putting personal 'fall protection' measures in place.

When choosing fall protection measures, collective measures such as guard rails or nets should be given priority over personal measures such as safety harness, etc. Whether it is personal or collective fall protection equipment, all equipment is subject to visual or more rigorous inspections and tests by appropriate persons. In all cases, test certificates will be required to ensure completeness of records. Workers have a right to see this certificate to check the last date of inspection or test.

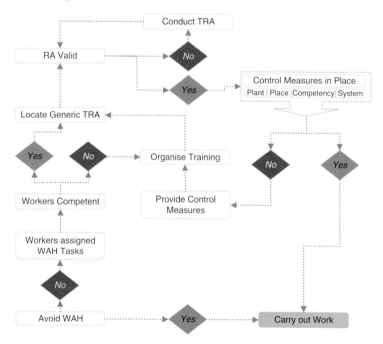

Figure 4.11 Basic WAH procedure

4.7.1 WAH Collective and Personal Control Measures

As discussed in the previous section, the risk of fall from a height can be mitigated through the use of collective or personal control measures. Examples of the equipment available to implement these control measures are briefly discussed here.

When considering controls for risks associated with WAH, the following hierarchy should be followed (Figure 4.12):

1. Avoid work at height by finding alternatives such as using telescopic extension equipment.
2. If avoiding WAH is not possible, then fall prevention measures should be considered giving priority to collective measures supplemented by personal measures. These measures may include work platforms, scaffolding or rope access.
3. Improve workplace to reduce the fall height or the impact of fall through additional control measures by utilising equipment.

4.7.2 Fall Restraint vs Fall Arrest System

Fall restraint systems prevent the wearers from falling whereas fall arrest systems stop the wearers after a fall and prevent them from hitting the object underneath by stopping the fall.

Examples of fall restraint systems include work positioning systems consisting of safety belts or full body harness secured

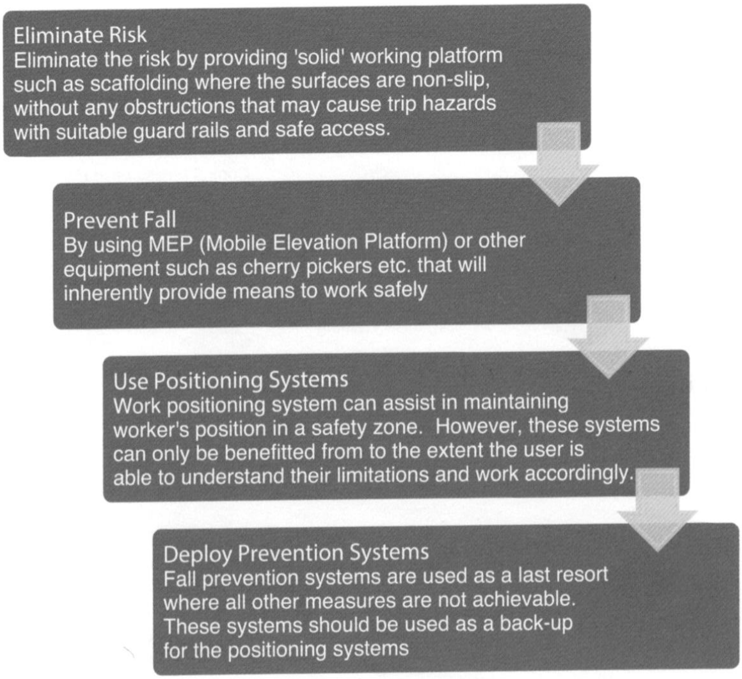

Eliminate Risk
Eliminate the risk by providing 'solid' working platform such as scaffolding where the surfaces are non-slip, without any obstructions that may cause trip hazards with suitable guard rails and safe access.

Prevent Fall
By using MEP (Mobile Elevation Platform) or other equipment such as cherry pickers etc. that will inherently provide means to work safely

Use Positioning Systems
Work positioning system can assist in maintaining worker's position in a safety zone. However, these systems can only be benefitted from to the extent the user is able to understand their limitations and work accordingly.

Deploy Prevention Systems
Fall prevention systems are used as a last resort where all other measures are not achievable. These systems should be used as a back-up for the positioning systems

Figure 4.12 Hierarchy of controls for working at height

around the body of the worker with the other end of these devices attached to a strong anchor point.

WAH equipment can be divided into the following categories:

1. Work positioning systems – use safety belt or full body harness leaving both hands free to carry out the task. Generally these systems are considered as 'personal' fall protection systems where the equipment is used to provide a supported position for carrying out the work. Fall arrest systems can be used to supplement the work positioning systems which prevent the wearer from hitting the ground or any other object at a predetermined height or falling in water when working over the side. Fall arrest systems can be fitted with energy-absorbing mechanism that will dampen the jerk when falling.
2. Travel restriction systems – these are the fall protection equipment that prevent fall, particularly when working close to the edge of a workplace.
3. Work restraint system – similar to the travel restriction system but attached to a fixed anchor point ensuring that the worker is not able to move to a distance larger than what is allowed for by the restraint system.

Various components that form part of the above three categories of WAH equipment are given in Table 4.1.

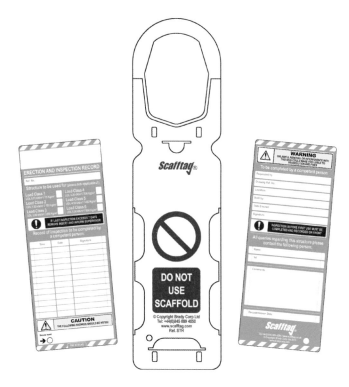

Figure 4.13 Safftag
(Source: www.scafftag.co.uk)

Table 4.1 – Working at height equipment

Equipment	Description	Equipment	Description
	Anchor Line Used for work on roofs or edges to prevent from falling over the side. Generally, these fall under the restraint and positioning systems.		**Head torch** An essential basic tool for working at night or in dark areas. Designed to fit with most of the helmets.
	Belt (Positioning) Adjustable belt used for work positioning. Can be used in combination with harnesses or other work positioning equipment.		**Helmet (Climbing)** Essential for head protection for work activities where a risk of fall exists such as rope access, work on towers, climbing or rigging.
	Ascender Used for ascending during rope access and climbing. These devices provide safety against fall prevention.		**Karabiner** Various types of karabiner connectors are used for various tasks.
	Descender Various types of descenders, for example, tower rescue and evacuation descender are to accommodate weight of one or two persons.		**Lanyard (Adjustable)** Various types of lanyards are available in the industry. The particular type shown here is used for work positioning on masts, pylons, towers or other lattice structures.
	Fall Arrester Various types of fall arresters are used in the industry. These are deigned to 'arrest' fall from a height and are attached between the harness and the anchor point.		**Lanyard (Twin Suspension Relief)** Similar to twin lanyard, but fitted with a foot loop providing a suitable location for a suspended worker to place foot.
	Foam Gloves All working at height tasks will require climbing to the work deck. These gloves reduce the wear on workers hands when climbing or handling ropes.		**Lanyard (Twin)** Used when moving between two work at height locations through provision of continuous connection to anchor points.
	Foot Loop Made of abrasion resistant material and reinforced to provide additional strength, these can be used when ascending or descending a rope or aid climbing in rope access.		**Nylon Sling** Used for various purposes e.g. extending lengths of lanyards, as an anchor point around a beam. Canvas rope protectors can be used around chaffing edges to provide extra safety.
	Harness (Climbing) Specially designed with front attachment point when working at height and rear attachment point for rescue use.		**Safety Glasses** For any task where eye protection is required, safety glasses of approved type must be used.
	Harness (Overhead Lines) Fitted with front and rear attachment point, usually adjustable for comfortable fit and fitted with leg loops, padded shoulder, waist and leg straps/loops.		**Waistcoat (High Vis)** Mostly yellow in colour with retro-reflective tapes. Orange coloured waistcoats are also used in the industry.
	Harness (Rope Access) Contains padded shoulder, waist and leg straps for wearer's comfort. Other accessories fitted to attach lanyards, ascenders/descenders etc.		**Harness (Evacuation)** Special harnesses for evacuation of personnel from height or even for vertical lift from underground workplaces.

The factors below should be considered for all areas of WAH:

1. Scaffolding should be erected by suitable qualified scaffolders who must only use approved equipment rigged in line with the standard 'BS EN 12811 – Temporary works equipment scaffolds: Performance requirements and general design'. This standard specifies the requirements for design of the access as well as working scaffolds. Duty holders must ensure that the safety system in place such as Scafftag is used appropriately including authorisation to install, alter or dismantle scaffolding.

2. Work on fragile surfaces can lead to falls through them resulting in death or serious injury. These incidents can easily be prevented through raising awareness of the issues, providing suitable equipment and following a safe system of work.

3. Personnel working at height would have mitigated all risks to themselves but the risks developed to others as a result of their work should also be considered. These workers will invariably be carrying tools and other work equipment to undertake their tasks; dangers to others from dropping these tools cannot be ignored. Measures should be taken to prevent tools or equipment falling down by fitting appropriate barriers. Areas in which overhead work is being undertaken should be cordoned off and personnel alerted of the dangers of falling objects. Tools should be carried in suitably designed carriers or belts to keep them secure during all work positions. In some cases, the tools may appear secure due to the barriers installed or raised edges of the work surface but the situation may change due to a change in circumstances such as weather, vibration or movement of the platform, especially on floating structures. Grated floors pose a specific risk as the tools may easily slip through openings of the grating.

4. Body posture has a very significant contribution to the avoidance of an injury, whether working on ground or at a height. However, it becomes even more important when working at height due to limitations posed by any accessories such as those discussed in this section. Care should be taken when a change in body posture stretches these accessories to limits affecting the ergonomical conditions of the workplace. Abnormal postures resulting from stretching body to reach areas further away from body or hanging over railings, etc. should be avoided to prevent risking an injury.

5. Use of ladders and step ladders is beneficial for short duration tasks that do not carry considerable working at height risks. The choice of ladder cannot be determined on the basis of the amount of time required for a task. However, where other measures are not practicable, task duration can provide a basic guide in decision making. For example, if a task requires awkward body posture for a continuous period of over 30 minutes, then

alternatives to the use of a step ladder should be considered. A thorough inspection of all parts of the ladder prior to its use is a must in addition to considering necessary precautions of carriage of tools whilst climbing the ladder, possibilities of over-reaching, the angle at which the ladder is secured (not more than 75 degrees) and obstructions in the vicinity of the work area. If a ladder is only employed for access to a platform, then it should be rigged in such a way that it provides at least 1 m height above the platform.

6. Mobile Elevating Work Platforms (MEWPs) can provide a safe alternative to portable ladders. The benefits of using MEWPs include providing access to locations at height as well as a working platform. However, their use requires more skill to operate than portable ladders. MEWPs should only be used by trained and authorised personnel. Additional use of work positioning systems is highly recommended to prevent risks associated with movement of the MEWP. All working at height precautions remain applicable even when MEWPs are being used to assist in any task.

4.7.3 WAH Emergency and Evacuation Plan

Like any other task being undertaken, contingency planning should be considered part of the initial assessment for undertaking working at height tasks. WAH regulations require employers to undertake this planning so that if things go wrong, any casualties can be attended to swiftly and recovered quickly. Whilst workers may be satisfied that the work positioning or restraint systems may be sufficient to prevent a fall, the impact of suspension trauma due to hanging in a harness is not commonly understood.

It is important that each worker understands the difference between 'rescue' and evacuation and the associated procedures.

Rescue is an event in which a casualty is recovered by other personnel whereas evacuation is when the control measures in place are breached and the workers themselves need to abort the task and find a safe exit. For each task being undertaken, consideration should therefore be given to all aspects of rescue and evacuation including:

- safety of the rescuers or evacuees;
- the time it will take to rescue workers;
- external assistance available such as the use of emergency services;
- speed at which assistance can be provided and the level of care available;
- skill of the rescuers;
- equipment available for rescue operation;
- worker's awareness for:

- contingency measures;
- details of the procedure to call for help;
- knowledge of equipment that may be used for evacuation or rescue;
- knowledge of limits whose breach will call for aborting a task and instigating the contingency plan.

Medical complications could easily result if a casualty remains hanging in a full body harness. If an action is not taken within the first 10 minutes to change a motionless casualty's position from upright to slightly elevated lower limbs, then the flow of blood towards the lower part of the body may cause the casualty to become unconscious. If the casualty is able to move, then they should be encouraged to raise legs and adopt a 'sitting' posture. This can be achieved by making a knee loop from a piece of rope or tape. A healthy person is likely to suffer from initial trauma resulting from a fall within 5 minutes. If not handled properly, this person can lose consciousness within 10 minutes. A further delay could quite easily lead to death in around 30 minutes.

When the casualty is recovered from the harness, they should never be laid flat since the pool of blood collected in the legs would not have circulated through the lungs and hence will have reduced oxygen level and increased carbon dioxide and toxic chemicals rendering it poisonous for the body. As the person is laid flat, this 'poisonous' blood will rush to the remaining circulatory system damaging all organs it passes through. If the casualty is laid down immediately after removal from harness, death may occur within about 10–20 seconds or even if the person survived, vital organs of the body could be damaged beyond repair.

In order to avoid the above eventualities, the casualty held vertically in a harness for 5–10 minutes or 30 minutes in a sitting position, should be lowered in a sitting position and kept in the same position for 30 minutes at least. During this time, the 'poisonous' blood will be released slowly back into the circulation system where it can regain its required level of oxygen and rid the chemicals and carbon dioxide. Expert medical advice should be sought subsequently for any further actions.

4.8 Mechanical Lifting

Accidents continue to happen during lifting operations on offshore installations due to various reasons including carelessness and incompetency of the workers, unsuitable equipment or inadequacy of control measures. The objects being lifted can vary significantly in weight, size and shape making lift operations complicated in many cases and therefore requiring specialised equipment as well as competency. When things go wrong, the damage could be caused

Figure 4.14 Scissor lift

to any entities involved such as personnel, installation itself, lifting equipment or the load.

In the UK, Lifting Operations and Lifting Equipment Regulations 1998 (LOLER) require that the lifting operations must be:

- planned in advance by a competent person;
- supervised at all stages;
- conducted safely.

The offshore oil and gas industry has produced several useful guidelines based on LOLER or similar regulations to facilitate common understanding of the underlying issues and procedures to overcome them. For example, the International Association of Oil and Gas Producers has published 'Lifting and hoisting safety recommended practice', UK's Step Change in Safety has produced a similar publication titled 'Lifting and mechanical handling guidelines'. These guidelines outline the essential principles for lifting and hoisting operations[2] as below:

1. Identify the lifting operation: This could be done by any worker who considers a 'lifting or hoisting' operation to pose some hazard that requires further assessment. Person(s) identifying the lift should inform the responsible person who will then nominate a competent person.

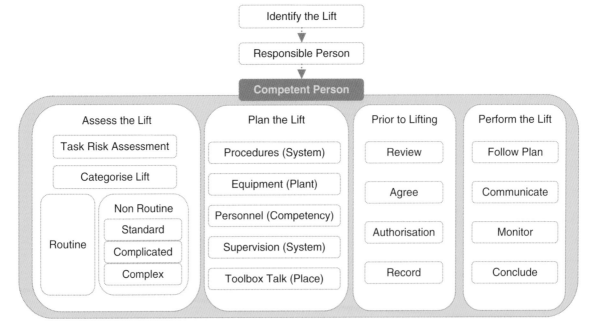

Figure 4.15 Lifting procedure

2. The competent person will carry out:
   a. Assessment of the lifting operation to determine its scope i.e. the details of the work to be carried out including risk assessment which should be conducted prior to commencing any work.
   b. As a result of the assessment, the lift can be categorised as:
      i. Routine lift – If a lifting operation is carried out regularly, then a generic risk assessment along with established safe procedures will be sufficient. However, it must be ensured that the control measures identified through generic risk assessment are in place. Most commonly, accidents happen due to complacency in compliance with procedures, ignoring the required controls or failing to take notice of changes in circumstances that may otherwise dictate a complete review of the risk assessment.
      ii. Non-routine lifts which can be divided into simple, standard lift, complicated or complex lifts. Sometimes a fourth category 'heavy' lifts is also included.
          A simple lift is where the object being lifted has a known weight and appropriate lifting points. The personnel involved in lifting the object are competent and the area through which the lift is being carried out is free from hazards.
          A standard lift is a simple lift but there are certain known and controlled hazards. The workers involved in this lift need to be aware of these hazards and the controls in place to conduct a safe lifting operation.
          The complicated or complex lift is where more than one crane is involved due to the weight, size, shape or location from where the object is being lifted or landed to. A lift where more than one object is being lifted simultaneously e.g. a number of tubulars being lifted as one lift also comes under this category. Heavy objects should come within this category and therefore same precautions should be applied.
   c. Plan the lift with details of:
      i. Procedures to be followed for preparation, conducting and concluding the lift.
      ii. Equipment to be used for the whole lifting operation. All equipment used must have appropriate approvals and certification which should be checked by the competent person. In addition, the competent person must inspect all equipment and accessories immediately prior to the commencement of the task. Any unsafe equipment should be quarantined and safe equipment clearly marked, for example, by using labels. The Safe Working Load (SWL) of the equipment should never be exceeded. An important factor to consider is the

availability of an uninterrupted power supply through-out the operation. A contingency plan must be included for power failure.

iii. Personnel assigned to the task should be assigned clearly stated roles. For example, successful communication between the crane operator and banksman is the key to successful completion of the task. A good understanding on the means of communication, particularly actions to take in emergencies, should be agreed and understood by both. A simple rule to keep in mind when assessing competency is to look for 'worker's ability to undertake a task safely'. Minor impact from drugs (medication) may hinder one's ability to undertake a safety critical job; hence all aspects of competency should be checked to ensure the ultimate level of safety.

iv. Supervision of the lift, preferably by the competent person. The role of supervision should be made clear to all personnel involved.

d. Perform the lift with the following steps:

i. Toolbox talk to discuss the plan, review it if necessary or agree if no further areas to be addressed are identified.

ii. Once the plan is agreed, it will need to be authorised by the responsible person. When authorised, the plan including all other documentation such as TRA becomes a record.

e. Conduct the lift following the plan. During this phase, the lift will be monitored under the safety observation system complying with all the requirements and stopping for any hazards as soon as they are identified. Once the task is complete, all equipment should be secured and personnel need to be rewarded even if it is simply by saying 'well done'.

Notes

1 HSE (2012) Manual Handling at Work: A Brief Guide. HSE Publication INDG 143. Available at www.hse.gov.uk/pubns/indg143. htm [Accessed 12.12.14].

2 OGP (2006) Lifting and Hoisting Safety Recommended Practice. Report No. 376 Published April 2006. Available at www.ogp.org.uk/pubs/376.pdf [Accessed 04.11.14].

5

Control of Substances Hazardous to Health

5.1 Hazardous Substances Offshore

Use of chemicals in any industry is part and parcel of most employees' work, especially in the oil and gas industry. However, this use must not put workers' health at risk. As a result, the Control of Substances Hazardous to Health Regulations 2002 (SI 2002 No. 2677) deal with procedures and precautions necessary for hazardous substances used in any work environment.

These substances can be found in any of the forms listed below but excluding lead, asbestos and radioactive substances:

- chemicals;
- fumes;
- gases;
- biological agents;
- dust;
- mist;
- vapours.

Figure 5.1 COSHH locker

Whilst COSHH regulations do not cover radioactive substances, lead and asbestos, there are some additional abilities of these substances such as entering the body through contact with skin, inhalation or ingestion because of which some of the COSHH regulations will apply. For example, when NORM come in contact with drill bits, pipe work, casing or other equipment, they may be contaminated, exposing workers handling this equipment. Similarly, asbestos dust can cause problems because of which the same requirements will apply as for other COSHH materials.

COSHH precautions also apply to a mixture of two or more substances found in any of the above forms provided they can cause harm if they come in contact with any part of the body, through inhalation or ingestion. Substances such as paints, grease, lubrication oils, cleaning fluids, disinfectants etc. are all covered by COSHH regulations.

Various terminologies used with reference to COSHH are:

1. Workplace Exposure Limit (WEL): about 500 substances have been identified with WEL which are given in the HSE publication 'EH40 Workplace Exposure Limits'. This publication defines two types of WELs i.e. short-term (15 minutes) and long-term (8 hours) exposure limits. These durations indicate the amount of time in any 24-hour period for which a person can be exposed to the given concentration of any chemical without harm to health. For example, the long-term WEL for acetone is 500 ppm or 1210 mg/m^3 and the short-term WEL is 1500 ppm or 3620 mg/m^3.
2. Parts per million (ppm): unit to describe concentration of chemicals or other substances in a given material. For example, 1 part per million could be 1 milligram of something in a kilogram of material.
3. CHIP: Chemicals (Hazard Information and Packaging for Supply) Regulations 2009. CHIP regulations have been replaced by the European CLP Regulations because of which CHIP was revoked from 1 June 2015. These regulations require suppliers to identify hazards of chemicals they supply, give information about hazards and package the chemicals safely. CHIP chemical classifications include very toxic, toxic, harmful, irritant, corrosive, sensitising, carcinogenic, mutagenic or toxic to reproduction.
4. CLP Regulations: European Regulation (EC) No 1272/2008 on classification, labelling and packaging of substances and mixtures came into force in January 2009 and adopts the UN's Globally Harmonised System on classification and labelling of chemicals across the EU including the UK.
5. NORM (Naturally Occurring Radioactive Materials) – radioactive materials for which human activity such as oil and gas exploration increases the exposure of workers to radiation are known as NORM.

6. Health Surveillance: assessment of the state of health of an employee, as related to exposure to substances hazardous to health, and includes biological monitoring.

7. Mixture: with reference to COSHH/CHIP/CLP, a mixture or solution of two or more substances is referred to as a mixture.

8. Substance: a chemical material and its compounds in a natural state or produced through a manufacturing process with any additives to preserve its stability but excluding any solvents removal of which does not affect the stability of the chemical material.

9. Dust: also termed inhalable dust means any airborne material which is capable of entering the nose and mouth by breathing. These materials are usually below 75 micro metres in diameter and can remain suspended in air for some time. COSHH regulations apply for dusts of any kind present at concentrations greater than 10 mg/m^3 inhalable or 4 mg/m^3 respirable. It is recommended that the exposure is limited to below 1 mg/m^3 for respirable dust and below 5 mg/m^3 for inhalable dust until other safe limits are put in place by the duty holder. When inhaled, particles of respirable dust can penetrate through the natural defences of lungs and reach the alveoli where gas exchange takes place. Consequently, these particles will remain lodged in the lungs (alveoli) forever leading to serious ill health in the longer term.

10. As Low as Reasonably Practicable (ALARP) – this is another term similar to 'so far as is reasonably practicable' used in the industry. The main focus of this term is on the words 'reasonably practicable' which requires duty holders to assess each situation according to the circumstances and take measures to reduce the risks of harm from any hazards to 'ALARP'. In allowing the duty holders to be the judge on the spot, the term allows them to base their decisions on the best ordinary practice of the industry. Consequently, the duty holder must assess all the risks and consequences. This will then lead to deciding the efforts/costs (in terms of money, time and trouble) that need to be made to control those risks. If it is established that the costs are not proportionate to the benefits of controlling the risks, then the duty holder can limit the costs to a suitable 'affordability' level which would take the risk to ALARP level. However, any obvious risks cannot be 'overlooked' purely on the basis of costs after reducing the risks to ALARP levels.

In order to protect workers and others, HSE[1] recommends the eight precautionary steps given below:

1. Assess the risks.
2. Decide what precautions are needed.
3. Prevent or adequately control exposure.

4. Ensure that control measures are used and maintained.
5. Monitor the exposure.
6. Carry out appropriate health surveillance.
7. Prepare plans and procedures to deal with accidents, incidents and emergencies.
8. Ensure employees are properly informed, trained and supervised.

5.2 How Workers Can Be Exposed to Hazardous Substances

Effects of hazardous substances on the human body can be of two types:

* Chronic or long-term effect – usually irreversible and occurs due to repeated exposure to the same chemicals or contaminants.
* Acute or short-term effect – the effects will be visible within a short period of time such as about 15 minutes. However, continued exposure to the same chemicals will lead to chronic effects. .

The damage to the human body may initially be local but can then spread to other organs. The extent of this damage will be determined by many factors such as the characteristics of the chemical, particle size, duration of exposure and concentration of the chemical substance to which an individual has been exposed.

Offshore workers can be exposed to hazardous chemicals through any of the following means:

* Inhalation – the effective control measure to avoid inhalation should be to control emission into the air within which workers are required to work or live. Means should be adopted to aim for zero emissions where possible otherwise appropriate PPE or RPE must be used. Exposure monitoring devices are considered the best tool to prevent any incidents.
* Skin contact – in oil production, it is natural that workers' body parts may come in contact with oils used in machinery or the produced oil. These oils, however, are intrinsically hazardous and therefore contact with them must be avoided by use of appropriate procedures.
* Skin injection: through cuts, bruises or other injuries which cause damage to the skin.
* Swallowing: chemicals or their fumes may be swallowed when mixed with other food products or water. Catering areas must therefore be kept isolated from hazardous chemicals at all times. Dirty clothing, gloves or other PPE which may have been

contaminated must not be allowed in dining areas. Hands must be washed prior to consuming any food.

- Damage to eyes – appropriate eye protection must be used to avoid damage to eyes. Workers must familiarise themselves with eye wash procedures and areas where eye wash kits are located.

5.2.1 Monitoring Exposure

The employer is responsible for monitoring the employees' exposure to any substance whose use may be considered hazardous under COSHH risk assessments. This is important as exposure is not always evident from symptoms of ill effect of the chemical substance being used. In many cases, the symptoms may appear months or even years after the exposure. If a substance in use has a defined WEL and if an employee has been over-exposed for any reason, then the employee must be taken through health surveillance checks to ensure integrity of their health.

Health surveillance may involve medical examination/surveillance, biological monitoring combined with inspection of COSHH records and medical history of individuals. For example, in cases where Local Exhaust Ventilation (LEV) has been identified as a control measure, it needs regular checking and thorough testing at least once every 14 months. However, any malfunction of LEV during a task will require all workers to be assessed for exposure. This process will not only require individual workers' health to be checked by medical surveillance and/or examination but also samples of the air to which they were exposed may also be tested, perhaps in a specially equipped laboratory. In some cases, additional measures such as biological monitoring may also be required through blood, breath or urine sampling. This will determine the quantity of chemical that has entered the body, particularly if the intake may have been by absorption through skin. This may also be applicable when an individual's PPE or RPE has been breached for any reasons.

In many workplaces, simple monitors are fitted to test samples of air in which workers breathe. These monitors will raise an alarm as soon as they detect the presence of hazardous chemicals in the atmosphere above a predetermined limit.

Monitoring may also be used to ensure any required control measures are operational and effective. Usually when employees are asked to participate in biological monitoring, they are required to give a written consent to the employer to show their agreement about:

- awareness of the actions that can be taken on the basis of results;
- who will access the test results;
- the substances for which they may be tested as the ones they may have been exposed to at work;

- the results of the test will not impact upon their rights as an employee.

It is recommended that all employees understand fully the consequences of participating in biological monitoring. Any doubts must be resolved prior to giving consent to the employer.

5.2.2 Sensitisation and the Difficulties of Monitoring Its Effect

Many substances cause allergy when they are inhaled or come into contact with the skin. Such substances are known as 'sensitisers' and can cause permanent damage to the skin, nose, throat or lungs. Respiratory sensitisers bring on hypersensitivity of the airway subsequent to inhalation of substance whereas skin sensitisers induce allergic response following substance contact with skin.

Symptoms for respiratory sensitisation include runny nose and/ or watery eyes with severe itching. If these symptoms persist due to the severity of the incident, asthmatic symptoms will appear such as tightness of chest, coughing and breathlessness.

Symptoms for skin sensitisation include dry, itchy skin which will change colour to red with possible cracking in the skin. If not dealt with properly, it may lead to bleeding and may even spread to other body parts.

Any of the above symptoms may appear on the initial short-term exposure. Mostly these appear after work because of which many workers will not relate symptoms to exposure at the workplace. If the reasons leading to these symptoms are not established and steps are not taken to avoid further exposure, then more serious consequences will follow. In many cases, symptoms appear months or years after the exposure. This is why any breaches of integrity of system must be reported and investigated immediately.

Employers or duty holders are required to carry out health surveillance of the employees. Examples of such surveillance are given in Table 5.1.

In cases with no obvious link between the illness and the workplace chemical hazards, the worker's record of sickness and any symptoms may prove vital for diagnosis by a medical practitioner.

Table 5.1 – Health surveillance tests	
Type of Chemical	**Test for Health Surveillance**
Respiratory sensitisers	Lung function test (Spirometry) to assess effect of exposure to chemicals
Skin sensitisers	Systematic observation of worker's skin by offshore medic

5.2.3 Hazard Symbols and Common Offshore Examples

Chemical suppliers are required to label packages with prominent 'symbols' according to the CHIP/CLP classifications below.

1. Very toxic
2. Toxic
3. Harmful
4. Corrosive
5. Irritant
6. Mutagen
7. Teratogen
8. Carcinogen
9. Sensitiser
10. Radioactive
11. Biological agent.

Figure 5.2 A worker handling chemicals

Some chemicals are only required to show actions required when things go wrong, for example spillage, but other chemicals require more information as the damage caused by them may be much more serious without any specified precautions and procedures for dealing with spillages or other incidents.

Where hazardous chemicals are supplied in bulk, for example, transferred from a road tanker to a reservoir, placing warning symbols may become impracticable. In such cases, the supplier must ensure the required information is supplied in the best possible way.

Hazard symbols are generally referred to as 'pictograms'. The meaning of these pictograms including any 'labelling' that goes with the pictogram is set out in law and therefore suppliers strictly follow the requirements. Where a substance or mixture is classified for more than one hazard, several pictograms may be used with the most severe hazard given a preference for inclusion in labelling. For example, for a package containing explosives, only the 'explosive' pictogram may be sufficient but additional symbols for 'flammable' and 'oxidising' substance may be used.

5.2.4 Reading a Material Safety Data Sheet (MSDS)

In Europe including the UK, a system called REACH (Registration, Evaluation, Authorisation and Restriction of Chemicals) for controlling chemicals is in use. REACH now incorporates the labelling and packaging requirements of CLP regulations. This became a law in the UK in 2007 where suppliers are required to supply MSDS or Safety Data Sheets (SDS) at the time of the first delivery of the hazardous substance or mixture.

The suppliers are required to supply SDS:

Table 5.2 – CHIP and CLP symbols

CHIP Symbols	CLP Symbols	Title	Meaning
		Corrosive	Corrosive materials that can damage metals, cause severe eye or skin damage
		Environmental Hazard	These substances are known to cause damage to aquatic environment
		Explosive	Self-reactive substances that may cause explosion
		Flammable	Flammable gases, liquids or solids which may be self-reactive, self-heating and can also emit flammable gases in contact with water
		Irritant/Harmful	Under CHIP, symbol X with letter i or h showing irritant or harmful substance. The new symbol under CLP regulations show less serious health hazard such as skin sensitisation
		Oxidising	Oxidising gases, liquids or solids which can produce a large amount of heat when they come in contact with other substances
		Toxic	Chemicals that can cause damage to health even at low concentrations
No current symbol		Health Hazard	New symbol under CLP regulations showing serious long-term health hazard for carcinogenicity and respiratory sensitisation
No current symbol		Gas under pressure	New symbol under CLP regulations showing compressed gas, liquids and solids, liquefied gases or dissolved gases

Table 5.3 – Hazard symbol categories

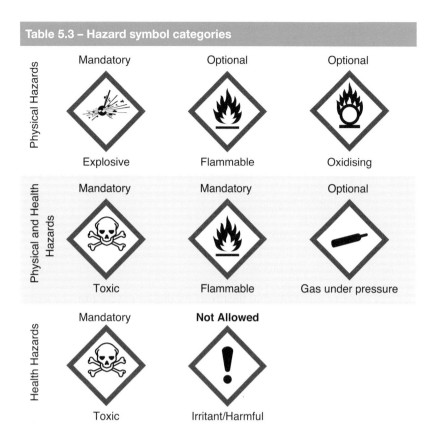

Physical Hazards		
Mandatory	Optional	Optional
Explosive	Flammable	Oxidising

Physical and Health Hazards		
Mandatory	Mandatory	Optional
Toxic	Flammable	Gas under pressure

Health Hazards	
Mandatory	**Not Allowed**
Toxic	Irritant/Harmful

1. free of charge, electronically or paper copy.
2. before the delivery of the substance or mixture. Availability of information from a website alone is not considered sufficient.
3. when they supply a substance or mixture that is:
   a. classified as dangerous under CLP Regulations or
   b. persistent, bioaccumulative and toxic or
   c. considered as a substance of very high concern.
4. when the customer requests an SDS for a mixture that is not classified as dangerous but can pose a hazard to human health or the environment or has Europe-wide workplace exposure limits.
5. when the product being supplied is listed as a special case with labelling derogations such as gas containers intended for petroleum gas, etc.

SDS must contain the following information:

1. date on which SDS was produced;
2. identification of the
   a. manufacturer/supplier;
   b. substance/mixture;
   c. hazards;

3. composition of substance/mixture or information on ingredients;
4. requirements for:
   a. first aid;
   b. fire fighting;
   c. accidental release;
   d. handling and storage;
   e. exposure controls/personal protection;
5. physical and chemical properties;
6. stability and reactivity;
7. information about the following for the substance/mixture:
   a. toxicology;
   b. ecological impact;
   c. transport;
   d. regulatory information;
   e. other relevant information;
8. disposal considerations.

It is important to note that the product labels must contain the name and contact details of the supplier, chemical and trade name of the substance or mixture along with risk phrases (R) or safety phrases (S). These are denoted as shown in the examples below.

Risk phrases specify special risks attributed to hazardous substances and materials. Examples of some risk phrases are given below:

R1	Explosive when dry
R7	May cause fire
R23	Toxic by inhalation

Safety phrases state the specific advice for hazardous substances and materials. Examples of some safety phrases are given below:

S1	Keep locked up
S(1/2)	Keep locked up and out of the reach of children
S21	When using, do not smoke
S29	Do not empty into drains

Further information about R and S phrases can be found from the HSE website www.hse.gov.uk or in the HSE publication 'Approved Classification and Labelling Guide'.

5.2.5 Employers' Duties under COSHH

Employers are responsible to control substances or mixtures that pose a threat to the health and safety of their employees. Under COSHH regulations, the following measures are required:

Figure 5.4 Workers fully kitted to work with chemicals

1. Assess the risks and establish control measures if prevention of exposure is not possible.
2. In establishing control measures, the following hierarchy of controls should be applied:
   a. Eliminate the hazard by redesigning the equipment or process.
   b. Substitute the hazardous substance by replacing the hazardous substance with the one that is non-hazardous or less hazardous.
   c. Engineer controls such as by using ventilation for extraction of fumes or smoke.
   d. Use administrative controls such as by reducing the probability of exposure to a hazard beyond WEL (working exposure limit).
   e. Use appropriate PPE or RPE.
   f. Prepare for emergencies by training employees in the correct use of hazardous substance as well as for dealing with any incidents or accidents.
3. Employers are required to report cases of dermatitis under Reporting of Injuries, Diseases and Dangerous Occurrences Regulations (RIDDOR) 1995.
4. Employers must establish a system of employees' health surveillance.

Employers are also required to ensure a system is in place which controls access to any chemical substances. Only authorised personnel should be allowed access to these substances to ensure unauthorised use does not create unnecessary hazards. All substances must be kept in secure lockers with suitable inventories showing quantities of each substance kept in storage.

Even when kept in storage, certain chemicals may pose some risks, particularly when kept alongside other chemicals which may react with each other. Hence the storage also requires a general risk assessment for all such chemicals which may react with others during their storage. Generally, the following must be kept separate:

• solid and liquid chemical;
• flammable and non-flammable liquids or solids;
• acids and alkalis;
• wastes from different chemical substances or materials.

5.2.6 COSHH Risk Assessments

Like any other tasks involving hazards and therefore requiring risk assessments, tasks where any chemicals are used must also be risk assessed, hence the name COSHH risk assessment or simply COSHH assessment.

COSHH assessment should not be simply taken as a paper exercise for compliance with legal requirements but must be seen as an exercise to find measures that can prevent exposure of workers to risks associated with the use of chemicals. Consequently, employers will be able to make decisions about prevention of the risks or establishing controls to at least reduce harm from those risks.

The basic process of COSHH risk assessment is the same as described in Chapter 2. However, in this case, competency of the assessor must be given paramount importance to ensure the person carrying out the assessment has adequate expertise of the hazards and risks associated with the substance being assessed in relation to the activity where the substance will be used. It may be prudent to team up the person with knowledge of the risk assessment procedures with the person(s) with adequate knowledge of the task and any substance being assessed. The person in charge of carrying out the risk assessment must have access to relevant legislation and ACoPS so that they can gather the required information. In cases where no 'expertise' is available on-board, the whole process may need to be outsourced to external organisations with suitable knowledge and experience.

Like any other risk assessment, a COSHH risk assessment is a live document as the risks should be based on the competencies of individuals carrying out the task. Furthermore, as the environment surrounding the task varies, the required control measures associated with the task may need to be varied requiring the risk assessment to be amended accordingly.

5.2.7 Employees' Duties under COSHH

Whilst employers have a duty to make provision for the health protection and safety of their employees, the latter also have a duty to protect their own and others' health and ensure safety. The Health and Safety at Work etc. Act (1974) requires employees to:

1. take reasonable care of their own health and safety and others who may be affected by their acts or omissions at work;
2. to cooperate with other persons in carrying out their duties so far as is necessary to enable that duty to be performed;
3. not to interfere with or misuse anything provided to ensure the health, safety or welfare of workers;
4. to use all work equipment and PPE appropriately;
5. to comply with the requirements of safe systems of work as set up by the employer;
6. to receive training as required by the employer;
7. to report defects in equipment and ensure control measures required by relevant risk assessments are kept in place;
8. to report any incidents/accidents related to or suspected of relating to use of chemicals/substances that come under COSHH procedures.

5.3 PPE Specific to Chemical Applications

For work areas where any chemicals are used, the associated risks will be determined by a risk assessment which would have considered the primary methods such as elimination, engineering or administrative controls. There may be cases where some residual risk needs to be controlled by the use of appropriate PPE. The selected PPE will be recorded in the relevant risk assessment documentation. As long as this PPE is used, the identified risks will remain controlled failing which the workers will be exposed to the hazards.

In order to ensure all risks are suitably avoided, the following factors must be considered:

1. For all hazards identified through risk assessment, select appropriate PPE as a final resort where residual risks can't be reduced further.
2. The employer is responsible for the provision of suitable PPE.
3. The employee has a duty to use the supplied PPE.
4. The employer must ensure employees are aware of all hazards including training in the PPE use, integrity checking and defect reporting.
5. Ensure PPE is maintained in good order and remains suitable throughout its period of use.
6. RPE (Respiratory Protection Equipment) is a special type of PPE used for protection against inhalation of harmful gases, dust, mist or fumes. Operations e.g. cutting a material such as stone (in drilling) or use of liquids that contain volatile solvents require the use of RPE.

Figure 5.5 A worker in full PPE

7. RPE is needed:
   a. where workers are at a risk of exposure to hazardous chemicals; the exposure may be through any means such as contact with skin, inhalation or ingestion;
   b. where there is a potential that fumes from a chemical being used may leak;
   c. in all areas exposed to chemicals or enclosed spaces where workers may need to respond to an emergency.
   Further details of RPE are provided in the next subsection.
8. Body protection is needed where:
   a. a chemical may spill posing a risk of coming in contact with skin;
   b. the chemical being used may catch fire.
9. Hearing protection is needed where:
   a. the noise level is high at all times such as machinery spaces;
   b. the noise level may suddenly increase due to certain equipment or machinery being switched on automatically such as compressor or pump rooms;
   c. due to existing noise, the workers need to raise their voice above their normal level to make the others heard at a distance of about 2 m.
10. Face protection is needed where:
    a. the reaction between chemicals may cause explosion;
    b. the chemicals include corrosives which may splash;
    c. there is a possibility of a splash fire.
11. Eye protection is needed where:
    a. there is a potential for a splash of a chemical;
    b. the temperature of a space is higher than normal.
12. In addition to the above, head and feet protection should also be considered with respect to the chemicals being used. Normal safety boots or hard hats may not be suitable or compatible with any specific PPE used for chemical hazards and therefore may easily be overlooked by workers. Special safety boots that resist the given impact, compression or penetration are available which may be coated with chemical resistant material.
13. When any PPE is supplied, it will come with manufacturers' instructions for its use, any limitations as well as storage. Many items are disposable i.e. one use only after which they must be disposed off securely following the instructions supplied and avoiding mixing items contaminated by different chemicals.

5.3.1 Types of RPE

The selection of RPE is critical to its effectiveness for workplace use. For example, certain atmospheres may contain more than one substance in the workplace that requires use of different RPE for each type of substance. However, it is practically not possible to wear two masks to protect against each substance. On this basis,

RPE is divided into the following types:

1. Filtering devices known as atmospheric respirators. These remove contaminants from the air being breathed in. Non-powered respirators are passive in the sense that they depend upon the wearer's ability to breathe in through the filter. Powered respirators on the other hand use motors/pumps to push air through the filter to enable breathing from clean air.
2. Breathing apparatus, also referred to as Self Contained Breathing Apparatus (SCBA) uses an air cylinder to supply air to the wearer.
3. Emergency Escape Breathing Device (EEBD) or ELSA (Emergency Life Support Apparatus). Both EEBD and ELSA provide a positive pressure on demand air for situations where escape requires exertion through areas filled with harmful gases or smoke. These come with compressed air sufficient for 10 to 20 minutes duration and should only be used for escaping in emergencies.

Figure 5.6 A worker in SCBA

RPE filters are a key component for correct breathing through the equipment. Various types of filters are available for different chemicals such as liquid/solid particles, gases or vapours. On the basis of the substances likely to be encountered in the atmosphere, filters are divided into the following types:

1. Particle filters – for airborne mixtures (solids or liquids) of mist or spray containing particles. These filters will not provide protection against gas or vapour.
2. Gas or vapour filters;
3. Combined filters.

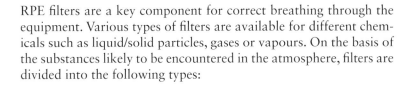

Figure 5.7 A worker in RPE and other PPE

All filters must only be used where a sufficient supply of oxygen is available. In some cases where various combinations of substances are present in the atmosphere, combination filters may be needed. However, the protection offered by filters will depend upon the protection factor of the filter. This is defined as a ratio between the hazardous substances outside of the filter/RPE in comparison with the quantity of the same substance inside the filter after having passed through it.

On the basis of protection factor, RPEs are categorised with an Assigned Protection Factor (APF). An APF of 10 will reduce the exposure to a hazardous substance by a factor of 10 which means one-tenth of the substance can go through the filter. Other examples of APF values used for filters are 4, 20, 40, 200 and 2000. Relevant Material Safety Data Sheets for hazardous substance provide these values. Keeping these values in mind, additional measures such as restricting the exposure to ALARP may also need to be considered for workers regularly exposed to hazardous substances.

The list below gives some basic types of the RPE available:
Respirators:

1. Disposable half mask – particle filter/vapour filter
2. Reusable half mask – particle filter/vapour filter
3. Full face mask – particle filter/gas/vapour filter
4. Powered mask/hood/helmet.

Breathing Apparatus:

1. Fresh air hose breathing apparatus
2. Constant flow airline respirator
3. Lung demand valve.

Particle filters used in RPE are marked with P and a number showing the filtration efficiency such as:

• P1 – low efficiency suitable for use with PF4 type respirators
• P2 – medium efficiency for use with PF10 type respirators
• P3 – high efficiency for use with PF20/PF40 respirators.

Figure 5.8 Disposable half mask (left); reusable half mask (right)

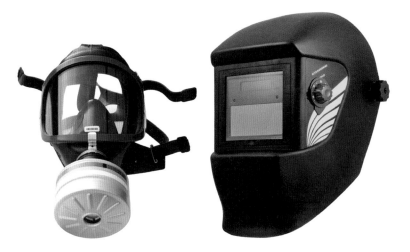

Figure 5.9 Full face mask (left), powered mask (right)

Further subdivision of filters is based on provision of fan-assisted filter protection for half and full masks (TM) or hoods, visors and helmets (TH). The particle filters P1, P2 and P3 can be used with fan-assisted respirators and therefore will carry a sign TM or TH as appropriate.

Gas filters are classified according the contaminant they can hold. These are:

Figure 5.10 A respirator filter

- Class 1: Low capacity
- Class 2: Medium capacity
- Class 3: High capacity.

Filters must always be changed according to the manufacturer's recommendations. Some may require to be changed after an hour's use; other may be used for a week before a replacement is required.

Table 5.4 – Gas mask colour codes	
Gas Mask Colour	**Used for**
Blue with white strip	Oxides of nitrogen
Brown	Acid gases, organic vapours with boiling point above 65°C
Green	Ammonia vapours
Grey	Inorganic gases and vapours
Red with white strip	Mercury
Violet	As specified by the manufacturer
White	Particles
Yellow	SO_2 and other acid gases

Figure 5.11 A trainee donning SCBA under supervision (Source: Ceronav Romania)

Gas filters are colour coded to provide easy recognition. These are detailed in Table 5.4.

5.3.2 Precautions when using RPE

The following factors must be considered for all circumstances where RPE is used:

1. Duration for which RPE is to be used must be considered. For example, for all cases of tightly fitted RPE, it should not be used for more than one hour to avoid causing discomfort to the wearer.
2. The suitability of the mask should be considered with reference to the size and shape of their face, presence of beard/moustaches on their face and whether they wear spectacles or contact lenses. It is recommended to carry out a 'fit test' with all wearers prior to putting them in an environment containing hazardous substances.
3. Wearers' physical health and any existing medical conditions should be considered to avoid aggravation of their condition.
4. Compatibility between the RPE and other PPE.
5. When RPE is in use, a procedure should be agreed to report any faults and abort operations if faults occur during the task.
6. For twin filter masks, both filters must be identical.
7. RPE must never be used in oxygen-deficient atmospheres.
8. All RPE must be cleaned and filters changed after use according to the manufacturer's instructions.
9. Damaged or dirty RPE must never be used. The filters are mostly one time use only and must be disposed off appropriately after use. Packaging seals of the new filters must be checked prior to use and expiry date checked. All filters with damaged seals must not be used.
10. For powered filters, stop work immediately if the fan/power stops.
11. If it becomes difficult to breathe through a filter, replace them immediately as they may have been choked due to high concentrations of dust or other contaminants.

Further information about the RPE can be found from the HSE publication *Respiratory Protective Equipment at Work – A Practical Guide* (ISBN 978-0-7176-6454-2).

Note

1 HSE (2012) COSHH: A Brief Guide to the Regulations, publication INDG 136 rev 3. ISBN 0 7176 2982 1. Health and Safety Executive.

6

Fire Fighting and Self-Rescue

Fire is considered to be one of the major hazards in the offshore oil and gas industry which can result quite easily from the ignition of accidentally released hydrocarbons from the well or any associated pipelines. Any components may fail accidentally due to vibration, corrosion, poor workmanship or other chains of incidents leading to HCR. Then all that is required is the source of ignition to light up hydrocarbons resulting in a fire. This chapter looks at the basics of fire, how to prevent it and if fire does occur, then how to evacuate safely.

Figure 6.1 Deepwater Horizon fire (Source: Flickr)

6.1 Fire Prevention

The offshore oil and gas industry employs two types of barriers to prevent fires: active and passive fire protection barriers.

- Active Fire Protection (AFP) systems are designed to extinguish fire. This system can use fixed installations based on foam, water deluge or gaseous extinguishing mediums. Portable fire fighting equipment also falls within the active fire protection system. AFPs are susceptible to failure due to their very nature, for example, by water leakage.
- Passive Fire Protection (PFP) systems provide protection for workers and equipment. Generally, this system slows down the spread of fire and hence provides extra time to escape from the danger area. Fire resistant doors and structural boundaries particularly in passageways, accommodation and muster stations form part of this system ensuring safe escape routes in the event of a fire on-board. PFPs require little or no maintenance and can provide safety for up to two hours in the fire and blast wall areas.

Fires can be divided into many different types depending upon the fuel they burn. Their speed of spread, the temperature they can reach and other features are contributory factors to decide control measures that need to be in place to safeguard against damage. Typical scenarios involving hydrocarbon and other fires in the offshore oil and gas industry include:

- Cellulose fires based on burning of wood, paper and textiles. These fires can reach a temperature of 500°C within 5 minutes, increasing to a temperature of over 1100°C.
- Electrical fires involving electrical equipment such as transformers, cables and control panels.
- Pool fires result from ignition of liquid fuel in the form of vapours above the horizontal pool of hydrocarbons. Pool fires can reach a temperature of 1000° within about 5 minutes.
- Jet fires result from liquid or gaseous hydrocarbons leaking from a storage container under pressure, hence can be very turbulent. In some parts of industry, jet fires are also referred to as spray fires but mostly jet fires are associated with gas fuels whereas spray fires are associated with liquid fuel leakages.
- Flash fire is an intense and extremely hot fire resulting from the ignition of a mixture of flammable/combustible liquid/gas that only lasts for about 3 seconds. This is also known as a fire ball. Whilst this fire itself is extinguished very quickly, it may ignite other objects it comes in contact with and set off a chain reaction.

- Explosion is a sudden movement of pressure or a shockwave resulting from the ignition of vapour cloud formed by the accidental release of gas or liquid hydrocarbons.
- BLEVE (Boiling Liquid Expanding Vapour Explosion) occurs when a liquid contained in a pressurised container reaches a very high temperature. If the relief valves fitted to the container are unable to cope with the increase in temperature and therefore increased pressure, the container will collapse causing BLEVE in the presence of a source of ignition.

In the UK, Offshore Installations (Prevention of Fire and Explosion, and Emergency Response) Regulations 1995 (PFEER) define the requirements for prevention, detection, control and mitigation of risks that may result from fire on-board. Oil & Gas UK published *Fire and Explosion Guidance* in 2001. This guidance provides best practice rules for prevention of fires and explosions and protection of personnel from these events as well as the arrangements for responding to these events. However, in order to ensure personal safety and escape safely, one must understand the basic principles of fire fighting as well as fire prevention. The basic rule is simple i.e. 'most fires are preventable'. Important factors to consider for prevention of fire include:

1. Raise awareness. Most accident investigations in industry reveal that any fire could be prevented through appropriate safety measures. These measures include hazard identification and risk assessment of all work or living areas leading to control measures to prevent fire. More importantly, when these measures have been put in place, all personnel must be informed about risks, control measures and consequences of neglecting these measures. The risks should be reduced to ALARP (As Low as Reasonably Practicable). Subsequently, planning for dealing with emergencies must form part of workers' 'induction' alongside training in the use of equipment.
2. Housekeeping plays a vital role in prevention of fires. Fire can start when its three ingredients i.e. oxygen, heat (ignition) and fuel are present in one location. Remove one of these ingredients, and the risk of fire will not exist. Offshore workplaces use a large variety of materials. Keeping them in appropriate designated locations will avoid the availability of all ingredients in one place and therefore avoid fire. Examples of untidy housekeeping areas that may contribute to the start of a fire include oil spills, waste including paper and oil soaked rags, unsafe (use of) electrical equipment or tools, mishandling, incorrect disposal and spillage of chemicals, galleys and even tumble dryers in laundries. Obstructions blocking escape routes could be lethal when the emergency escape plan needs to be implemented. A good housekeeping routine where all workers participate

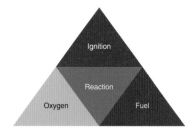

Figure 6.2 Fire triangle

actively therefore must establish general tidiness across all areas mentioned here.

6.2 Fire Science

Fire results from a combination of oxygen (or air), fuel and a source of ignition producing heat, smoke, light and toxic fumes. These three components are referred to as the sides of a triangle, referred to as the fire triangle. If any one side of the triangle is missing, then the triangle will be incomplete and the reaction that produces fire will not take place. The three components that form fire may become available from:

- Oxygen – available from air. Normally 21 per cent oxygen is available in the air but 16 per cent is considered sufficient to support fire.
- Fuels, which can be divided into two categories, flammable and combustible materials, the former having the ability to catch fire quickly with a small ignition source such as propane or butane gases whereas the latter have the ability to burn under specific conditions and will take longer to catch fire, for example, wood or paper, etc. Flammable materials burn very rapidly and therefore produce an explosion whereas combustible materials burn at a steady rate and give off a lot of smoke. Fuels can be in:
  - liquid forms such as grease, oil or liquid fuel oils;
  - solid forms such as paper, wood or even metals such as magnesium, zinc, titanium, potassium, etc.
  - gas forms such as hydrocarbon gases from the production or export areas of an oil field.
- Ignition source or heat, which can come from:
  - flame such as in a galley or other areas where open fires are used regularly;
  - spark from electric equipment;
  - hot work such as welding and cutting;
  - chemical reaction between various chemicals;
  - heat produced due to friction between various components of machinery;
  - pressurisation of gases;
  - spontaneous combustion.

In order for fire to take place, the fuel or combustible material must produce vapours that burn. For example, gasoline can catch fire quickly if compared with diesel oil because the former evaporates quickly in comparison with the latter. Once the oxygen, fuel and source of ignition combine, the resulting product (fire) can spread quickly through the spread of heat since the other two sources are available in almost all workplaces. The spread of heat could be through any combination of the following processes:

- Convection – hot and therefore lighter air/gas travels upwards causing vertical spread of fire. The same phenomena can take place within liquids.
- Conduction – movement of heat within materials or objects in direct contact with each other. The heat transfer rate through conduction will depend upon the material itself e.g. wood will have less conduction support for heat than steel.
- Radiation – transfer of heat through infrared rays where heat is transferred through empty space around the fire.

Regardless of how the fire starts and spreads, its by-products are CO_2 (carbon dioxide), CO (carbon monoxide), H_2O (water) and nitrogen gas in addition to smoke. Smoke is made up of carbon and other unburnt particles floating in the air in the form of tiny particles that may be visible to the naked eye. These particles may contain moisture produced as a result of the combustion process. The nature and content of other chemicals produced will depend upon the 'fuel' being burnt. When this mixture is inhaled, it could cause irritation or choking depending upon the quantity of smoke inhaled. It is estimated that 50–80 per cent of deaths in fire result from smoke inhalation.

The most dangerous of gases produced from combustion is CO which is produced as a result of incomplete combustion of fuel. Inhaling a mixture of CO and air can be a serious health hazard since CO can be absorbed in blood before oxygen causing deficiency in the oxygen required for the proper functioning of the brain and other parts of the body. When the human body is exposed to CO concentration of 1.3 per cent or above, this can cause unconsciousness within 2–3 breaths and death within minutes.

When CO_2 is present in higher concentrations than normal, the inhaled mixture of CO_2 and air will contain less oxygen causing breathing to be deep and rapid to meet the body's demand for oxygen not being met by the available level of less than 21 per cent in the air. At about 15 per cent oxygen, breathing becomes difficult, below 15 per cent fatigue sets in and judgement is impaired significantly with below 10 per cent oxygen causing unconsciousness.

In addition to the above problems caused by fire, other complications can arise as a result of loss of power and therefore inability to find escape routes, and injuries caused due to burns, etc.

6.3 Fire Terminology

Flash Point

Lowest temperature at which a liquid generates enough vapours to form a flammable mixture in the air.

Table 6.1 – Flash point and auto ignition temperature examples		
Product	**Flash Point**	**Auto Ignition Temp**
Acetylene	−18 °C	305 °C
Butane	−60 °C	405 °C
Methane	−180 °C	580 °C
Petrol (Gasoline)	−43 °C	280 °C
Propane	−104 °C	470 °C
Vegetable Oil	327 °C	360 °C

Auto Ignition Temperature (AIT)

Temperature at which a material ignites without an external source of ignition.

Spontaneous Combustion

Ignition of a substance without introduction of a source of heat or ignition.

Volatile

Materials are considered to be volatile if they vaporise easily at low temperatures and are prone to being ignited easily.

Vapour Density (VD)

Ratio between the weight of a vapour compared to an equal volume of dry air. This ratio varies with temperature. If this is greater than 1 (air), then the vapour is heavier than air and will settle in low places below air whereas if it is less than 1, it will rise above air. Some examples of vapour densities are given in Table 6.2.

Table 6.2 – Vapour density comparison			
Product	**VD**	**Air VD**	**VD**
Steam			0.80
Methane			0.67
Carbon Monoxide			0.97
Air		1.0	
H$_2$S	1.19		
Propane	1.52		
Petrol (Gasoline)	4.40		
Butane	2.00		
Chlorine	2.49		

Boiling Point	Temperature at which a liquid will change to vapour under normal atmospheric conditions.
Flammable Range	The range of a concentration of a gas or vapour at which it will ignite if a source of ignition is introduced. Also see LEL and UEL.
Combustible/Flammable Material	Flammable materials have the ability to catch fire quickly with a small ignition source such as propane or butane gas whereas combustible materials have the ability to burn under specific conditions and will take longer to catch fire, for example, wood or paper, etc.
Lower Explosive Limit (LEL)	The lowest concentration (expressed as a percentage) of explosive gas in an atmosphere at which fuel is not enough to continue explosion.
Upper Explosive Limit (UEL)	The highest concentration (expressed as a percentage) of explosive gas in an atmosphere at which oxygen level is not enough to continue explosion. Some examples of LEL and UEL are given in Table 6.3.

Table 6.3 – LEL and UEL examples		
Product	LEL (%)	UEL (%)
Acetylene	2.5	100
Butane	1.8	8.4
Methane	5.0	15
Petrol (Gasoline)	1.2	7.1
Propane	2.0	9.5

DSEAR	Dangerous Substances and Explosive Atmospheres Regulations 2002 (DSEAR) (SI 2002 No. 2776)
ATEX	Name given to the two European Directives for controlling explosive atmospheres in the workplace. These are implemented in the UK through DSEAR 2002:

- Directive 99/92/EC on minimum requirements for improving the health and safety protection of workers potentially at risk from explosive atmospheres.
- Directive 94/9/EC on the approximation of the laws of Member States concerning equipment and protective systems intended for use in potentially explosive atmospheres.

Classification of Zones

Hazardous places are classified as 'zones' based on the frequency and duration of the occurrence of an explosive atmosphere.

Zone 0: Explosive atmosphere where a mixture with air of gas, vapour or mist exists continuously, or for long periods or frequently.

Zone 1: Explosive atmosphere where a mixture with air of gas, vapour or mist is likely to occur in normal operation occasionally.

Zone 2: Explosive atmosphere where a mixture of gas, vapour or mist is not likely to occur in normal operations but, if it does occur, will persist for a short period only.

Zone 20: Explosive atmosphere where a cloud of combustible dust in air is present continuously, or for long periods or frequently.

Zone 21: Explosive atmosphere where a cloud of combustible dust in air is likely to occur in normal operation occasionally.

Zone 22: Explosive atmosphere where a cloud of combustible dust in air is not likely to occur in normal operation but, if it does occur, will persist for a short period only.

Structural Boundaries

Structural boundaries are formed by bulkheads and decks complying with the criteria shown in Table 6.4.

Table 6.4 – Structural boundary classification			
	A Class	**B Class**	**C Class**
Construction	Steel or other equivalent	Non-combustible materials	Non-combustible materials
Average temperature unexposed side	Not higher than 140°C above original temperature	Not higher than 140°C above original temperature	
Spot temperature	Not higher than 180°C above original temperature	Not higher than 225°C above original temperature	
Time limitations for subclasses	A-60: 60 minutes A-30: 30 minutes A-15: 15 minutes A-0: 0 minutes	B-15: 15 minutes B-0: 0 minutes	

6.4 Classes of Fire

Fires are classified by the type of 'fuel' they burn. Any fire extinguishing medium used to extinguish a fire depends upon the type of fuel involved in a fire. These are described below.

Class A	Fuel for this type of fire is ordinary combustible material such as paper, wood, cloth and plastics. A water extinguisher is considered best for Class A fires.
Class B	Fire caused due to burning of flammable liquids such as hydrocarbons (gasoline, diesel oil, etc.) or paints. Depletion of oxygen is the best way of controlling Class B fires, hence AFFF (Aqua Film Forming Foam) extinguishers can be considered best for flammable liquid fires.
Class C	Fire caused due to burning of flammable gases like propane, butane, etc. are classed as Class C fires.
Class D	Combustible metal fires fall within Class D fires. Examples of such metals include swarf, shavings or powder of potassium, titanium, aluminium, magnesium and sodium. Dry powder extinguishers are considered best for this class of fire since they absorb heat as well as aid smothering.
Class E	Electrical apparatus based fires come within Class E fires. In order to tackle a Class E fire, the source of electricity must be removed and then a CO_2 based extinguisher should be used.
Class F/K	Cooking oils, animal fat and grease fires are given a separate class instead of keeping them in Class B fires because conventional extinguishers are not considered effective due to the high temperatures (around 360°C) at which these 'fuel' materials ignite. Specialised wet chemical extinguishers have been developed to tackle these fires. This type of fire is classed as F in the EU and Class K in the USA.

6.5 Extinguishing Fires

The basic principle of extinguishing any fire remains the removal of one side of the fire triangle. This can be achieved by the following:

Cooling	As described previously, when the temperature of the fuel or burning material rises, it gives off vapours that burn. If the fuel is prevented from producing vapours then the fire can easily be extinguished. Water is the most effective medium to use on fires, especially Class A fires. It impacts in two ways, first by reducing the temperature of the fuel directly and second by cooling through evaporation. As the water gets heated, it evaporates causing a reduction in the temperature of the object on fire. However, the use of water in enclosed spaces should be carefully considered since some steam will be generated as a result of water coming in contact with hot surfaces. The volume of this steam is likely to reach around 1,700 times the volume of water used. Another consideration for the use of water is the impact on the stability of the platform or the vessel. If the use of water within a space is considered to cause steam related problems, boundary cooling should be considered. In this method, instead of applying water directly to the fire site, the boundary of the space on fire is sprayed with water to lower the temperature.

Starvation   Fire can be extinguished simply by removing the fuel that is burning and keeping the fire alive. Depending upon the situation, this can be achieved by switching off the electric, gas or fuel oil supply or other materials on fire and in addition removing flammable or combustible materials from the vicinity of the fire. It must be remembered to remove these materials from all six sides of the place where a fire exists. In addition, natural or mechanical ventilation should be closed to cause further starvation of oxygen in the subject space.

Smothering   In this method, the oxygen supply to the fire is removed by use of smothering substances or agents. For example, foam used on the surface of an oil fire not only cools the temperature down but also covers the surface of oil, thus removing the supply of oxygen and hence extinguishes the fire. Any inert gas or CO_2 can also be used to smother a fire.

Inhibiting   Some substances chemically react with materials on fire to break the chain reaction between the three sides of the fire triangle and thus extinguish the fire. These substances are known as flame inhibitors. An example is the use of dry chemical fire extinguishers.

6.6 Actions on Discovering Fire

Fires always start small, and then spread rapidly depending upon the materials that are on fire. The initial response to a fire will determine the final outcome. The following sequence of actions should be kept in mind when tackling a fire:

1. Stay calm and think about the actions taken during fire drills on-board.
2. Raise the alarm as soon as the fire is discovered. This does not necessarily mean activating the alarm which can be done when one gets to it but prior to activating the alarm, shout 'Fire, Fire, Fire...' until the person reaches the activation point.
3. Inform the control station about the location, nature and size of the fire, any injured persons, the damage that has been done and the actions that have been taken.
4. Ensure own safety first and then assess if it is possible to:
   a. Extinguish the fire by using a portable extinguisher. If not, then don't take risks and ensure own safety. If a portable extinguisher is not enough to extinguish the fire, then it requires other means beyond the ability of the fire being handled by just one person. In this case, proceed immediately to the muster station.
   b. Subsequently, if it safe to do so, restrict the spread of fire by closing doors/windows to cut the air supply, shut down the supply of electricity, fuel or gas, remove other combustible/flammable areas from the vicinity of the fire.
5. Do not use lifts after a fire alarm has been activated.
6. Do not go through automatically closing watertight doors when their closing mechanism has become operational to avoid being trapped.
7. Arrive at the muster station and wait for further instructions.

The above steps are summarised below for ease of remembrance:

Find	Fire could be detected either by a person or through smoke or fire detectors. All personnel must remain vigilant at all times to see the fire or smoke, smell burning or hear the alarm raised by others.
Inform	Inform the control station by raising the alarm immediately. If the alarm activation point is not in the immediate vicinity, then shout 'Fire, Fire, Fire...' loudly when going towards the activation point.
Restrict	Restrict the fire by turning off the electric/gas/fuel supply for the area, removing other combustible/flammable materials from the vicinity of the fire, closing down ventilation and shutting doors and windows.
Extinguish	Extinguish the fire if possible, for example a small fire in a waste bin, but if it is beyond your control, then evacuate the area and report to the muster station to let the fire party deal with it or to get extra help.
Escape	

6.7 Portable Fire Fighting Equipment

All workplaces are required by law to be supplied with portable fire fighting equipment. Portable extinguishers are the most common in all workplaces. These are spread around strategic locations in such a way that users don't have to go too far to find one.

In order to ensure ease of use, portable extinguishers are generally similar in their external appearance as well as the operation mechanism. For example, a dry chemical extinguisher releases the dry chemical stored inside the cylinder of the extinguisher through the siphon tube (Figure 6.3). The operation can be carried out simply by removing the safety pin and pressing on the discharge lever when the nozzle has been pointed in the required direction. Minor differences for other types of extinguishers such as water, foam, CO_2 and wet chemical extinguishers are due to their contents. Most modern extinguishers use CO_2 bottles inside the cylinder instead of stored pressure. When the discharge lever is operated, the CO_2 is released into the cylinder which pushes the cylinder contents out through the nozzle.

In the UK, the British Standards published by the British Standards Institution (BSI) control the design of portable fire extinguishers as given in Table 6.5.

General information about siting of fire extinguishers, labelling and colour coding is given below:

Figure 6.3 Activating a fire extinguisher

Table 6.5 – British Standards applicable to fire extinguishers	
British Standard (BS)	**Details**
EN3 Part 7 2004	Characteristics, performance requirements and test methods
5306 Part 3 2009	Commissioning and maintenance of portable fire extinguishers
5306 Part 1 2006	Fire extinguishing installations and equipment on premises
5306 Part 8 2007	Selection and positioning of portable fire extinguishers
6643 Part 1 2008	Recharging fire extinguishers
6643 Part 2 2008	Specifications for powder refills for extinguishers
7273 Part 1 2006	Operation of fire protection measures

FIRE SAFETY SET

Figure 6.4 Portable extinguisher colour coding (colour is for a band on the cylinder)

1. Portable extinguishers should not weigh more than 20 kg.
2. Extinguishers are positioned in such a way that their operating handle is 1m from the deck level for extinguishers heavier than 4 kg and 1.5 m for lighter extinguishers.
3. They should be sited close to the entrances/doors and away from extreme pressures. Special consideration should be given to fitting some extinguishers on escape routes.
4. Generally, they should be kept at distances as given below but specific requirements will be based on the layout of the place, availability of other fire extinguishing medium such as fixed installations, etc.
   a. Class A zones – 30 metres
   b. Class B zones – 10 metres

    c. Class C zones – 30 metres
    d. Class D zones – distance based on safety case
    e. Class F zones – 10 metres
    f. Accommodation spaces – generally within 15 metres.
5. Extinguishers should be marked with the following labels:
    a. the word 'Extinguisher';
    b. extinguishing medium such as water, foam, CO_2, etc.;
    c. types of fire for which an extinguisher is suitable;
    d. procedure for use of the extinguisher, preferably given in pictorial format;
    e. precautions, risks and dangers of use;
    f. any areas where extinguisher is unsuitable;

Table 6.6 – Fire extinguisher colour coding

Type	Colour (Band)	Fire Class	Method of Use
Water	Signal Red	A	Plain water is an excellent cooling agent and therefore extremely effective on Class A fires. When deploying an extinguisher, point the nozzle at the base of flame moving it to cover the entire area of fire until extinguished. Fire fighter should stay low to avoid steam and heat. Once the fire is out, use the full extinguisher to moisten the fuel (wood, paper, etc.) to prevent re-ignition.
Foam	Cream	A, B	Foam is produced by the use of protein based compounds or detergents mixed with water. Foam can float on the surface, hence point the nozzle at the base of flame moving it to cover the entire surface with foam until extinguished. Maintain some distance from fire depending upon the throw of extinguisher. Do not aim the jet into the liquid to avoid a splash which will spread the fire further. On a fire with enclosed boundaries such as a fuel drip tray, spray the foam on the sides and let it flow down to the surface of the liquid. Always fully discharge the extinguisher and keep back-up ready for use if required.
Dry Powder	French Blue	A, B, C	Dry powder is a fine chemical compound that acts through separating the three sides of the fire triangle. Point the nozzle at the base of the flame giving it a sweeping motion. This acts quickly and chemicals used do not lower the temperature of the burning material; hence always discharge the extinguisher fully.
CO_2	Black	A, B, E	CO_2 is a non-conductive gas that displaces oxygen to smother a fire. Point the horn/jet just above the base of flame moving it across the area to cover it entirely. CO_2 is discharged at a high pressure; hence avoid placing the nozzle too close to the burning material to avoid splashing/spreading it around. Pay particular attention to the freezing temperature of nozzle due to liquid CO_2 changing into gas as it comes out of the cylinder.
Wet Chemical	Canary Yellow	A, K	Point the extended applicator to the base of the flame and move in circular direction covering the surface of the burning fat/grease/cooking oil. No areas should be left empty to avoid re-ignition.
Special Powder	French Blue	D	These extinguishers are equipped with special low velocity applicator to apply the medium effectively by covering the surface of metal swarf or powder.

g. manufacturer's name and address;

h. in addition to the above, date of last service and inspection should also be given on a separate label attached to the extinguisher.

6. The body of all extinguishers is required to be RED in colour with an allowance of 5 per cent of the body to be colour coded according to the extinguishing agent contained in the cylinder as per the guidance given in Table 6.6.

6.7.1 Fire Blankets

Fire blankets are made from flame resistant material and are used to extinguish small fires in areas such as kitchens or galleys where pan fires are quite common. The blanket can also be used to wrap around a person whose clothing catches fire. Fire blankets work by cutting the oxygen supply to the fire, removing one side of the fire triangle and thus extinguishing it.

Fire blankets are housed in a casing mounted on the bulkhead/wall and are one use only. They can be pulled out of the casing using the attached straps which usually remain hanging outside of the casing. Induction for all employees should include familiarisation with the procedure for use of the fire blanket. Time is precious when one has to deal with an emergency and therefore no time should have to be wasted in reading the instructions. In order to use a fire blanket effectively, the recommended guidance is given below:

Figure 6.5 Fire blanket in galley

- Ensure the fuel supply such as gas is turned off and the area where the fire needs to be extinguished is clear of obstructions or other hazards.
- Pull the blanket out of the casing holding it from the straps provided.
- Wrap the top edges of the blanket around the hands to avoid burns.
- Lift the blanket as high and as wide as possible ensuring your view is not blocked and step towards the fire carefully.
- Ensure your escape route is clear.
- Cover the flame with the blanket starting at the top and moving the bottom towards the base of the fire. It is important to cover

Figure 6.6 Large trolley extinguisher

Table 6.7 – Large extinguisher capacities	
Medium	**Capacities**
Foam	50, 100 litre
Dry Powder	30, 50, 75, 100 Kg
CO$_2$	10, 20, 30, 45, 60 Kg

the whole flame so that the oxygen supply can be cut off without which the fire will not extinguish.

- Once the blanket covers the fire, leave it in place for some time (at least 30 minutes) to ensure the fire has been extinguished. Remember, cooking oils, grease and fats need to be heated up to over 300°C to catch fire. If the blanket is removed too quickly, oil will still be very hot and may re-ignite.

When a fire blanket has to be used for a person on fire, it should be wrapped around the person who should be laid down on the deck immediately after being wrapped the blanket. Once the fire is extinguished, the blanket must be removed immediately to avoid burns to the skin from burnt clothing.

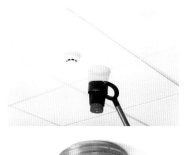

6.7.2 Large Trolley Extinguishers

Large areas requiring greater first aid fire fighting can be covered by large extinguishers placed on trolleys. Generally, these are located in machinery spaces, but since they are portable, they can be brought to location such as bunkering manifolds during fuel oil transfer operations.

Despite their size, the principle of operation for these extinguishers is exactly the same as the portable extinguishers. They are designed to be manoeuvred and used by one person. The quantity of the extinguishing medium in the cylinder is dictated by the size of the space. Table 6.7 gives some examples of common trolley extinguishers available in the industry.

Figure 6.7 Smoke detector with test kit (above), sprinkler head (below)

6.8 Fire Detection and Extinguishing Systems

Fixed fire detection and extinguishing systems are fitted to offshore installations to protect from specific risks. Considerations for the choice of system include:

- the time it will take to detect the fire and to extinguish it;
- the damage that is likely to be caused to the equipment within a space as a result of using the extinguishing medium;
- time and effort required to make the system operational after the event.

6.8.1 Fixed Fire Detection and Alarm Systems

Offshore installations are fitted with fire detection and alarm systems, commonly referred to as fire alarm systems, consisting of the following components.

1. fire detection devices such as heat or smoke detectors;
2. manual call points for activation of fire alarms;

3. fire alarms and associated warning lights;
4. control and indicator panel.

A conventional fire alarm system is based on division of the installation into zones with each zone connected to a control and indicator panel. Zones are created on the basis of fire protection boundaries.

Table 6.8 – Types of fire, smoke, heat detectors

Types of Detectors	Description
Heat Detectors	Detect increase in temperature due to fire.
Fixed temperature	When the temperature increases to a predetermined level, then this detector operates. Restorable fixed temperature detectors restore to their normal state after the temperature drops whereas the non-restorable heat detector element has to be replaced once it has been activated.
Rate of rise	If the rate of increase of temperature per second (or per minute) exceeds the predetermined set rate of temperature increase, then this detector operates.
Rate compensation	These detectors operate on the same principle as the rate of rise detectors but are fitted with rate compensation mechanism to adjust for thermal lag. These are used in areas where weather or moisture is likely to have an impact on the atmosphere around the detector or in areas which have explosive atmospheres.
Fibre optic	In these detectors, optical fibres made of quartz glass are used as sensors. Change in temperature can be more precisely monitored at a distance of many kilometres in comparison with other detection systems.
Flame Detectors	
Optical (infrared/ultraviolet)	These detectors detect the optical characteristics of the flame by use of one or more optical ultraviolet or infrared sensors to distinguish between the flame and background radiation. Optical detectors are considered superior to the traditional detectors due to their accuracy and ability to 'see' fire at a larger distance.
CCTV	Optical detectors can be combined with CCTV cameras to aid investigation of a false alarm at a distance.
Smoke Detectors	
Ionisation	The detector is fitted with a radioactive material between two metal plates connected to live electric connections. The radioactive material ionises the air between these two plates. Smoke can block this ionisation process, stopping the flow of current and thus triggering the alarm. Because of their mechanism, these detectors are considered more responsive to flaming fires.
Photoelectric (or optical)	These detectors use a directional light projected in a straight line within the detector chamber. As smoke enters the chamber, instead of continuing to be projected in a straight light, the light is deflected towards a sensor causing the alarm to activate. These detectors are considered more useful for areas where fire will smoulder for some time before it reaches to flame stage.
Video smoke detector	This system operates by processing the video image from a digital or analogue camera by using special algorithms using image processing through computers. These detectors can also be used for fire detection.

CO Detectors	Carbon monoxide (CO) is produced as a result of incomplete burning of fuel in any intentionally or unintentionally lit fire. CO detectors detect the presence of CO above a predetermined level. Various types of sensors are used to detect CO in atmosphere.
Bimimetic	Various chemicals that change colour when affected by CO are used. Flow of light through these chemicals can trigger an alarm.
Electrochemical	It uses a fuel cell that allows passage of electric current. Amount of current determines the percentage of CO in atmosphere. If it reaches above a predetermined level, an alarm is raised.
Semiconductor	These operate on a principle opposite to that used for electrochemical sensors i.e. a metal such as tin is connected between two components of a circuit. Air/oxygen increases resistance against flow of current whereas CO reduces this resistance triggering the alarm when it exceeds a given level.
Combined Detectors	Various detectors can be combined to detect smoke, fire, heat and CO to improve the detection system. For example combined optical and ionisation, CO and smoke are available in the market.

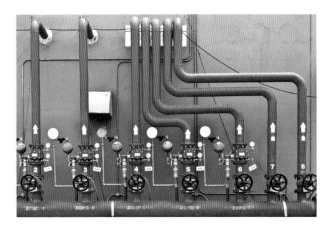

Figure 6.8 Water sprinkler control system

6.9 Fixed Fire Extinguishing Systems

There are normally two types of fire protection systems i.e. active and passive protection systems.

Active Fire Protection (AFP) systems include fixed fire extinguishing systems which will be activated in the case of activation of the fire alarm through detection of fire, smoke or heat. Passive Fire Protection (PFP) systems on the other hand are aimed at containing fires or reducing the spread of fire by use of fire resistant materials for construction of the installation. PFP requirements include the size of spaces, type and materials used for construction.

In the UK, each offshore installation's Fire and Explosion Strategy (FES) defines requirements for a particular installation. Offshore Installations (Safety Case) Regulations 2005 in the UK require all installations to evaluate major risks and put control measures in place for these risks. Offshore Installations (Prevention of Fire and Explosion, and Emergency Response) (PFEER) Regulations 1995 state the requirements for mitigation of fire and explosion as well as for musters and drills.

The AFPs are described in this section.

6.9.1 Sprinkler Systems

Various types of water based fixed fire fighting systems are used as sprinkler systems. These include:

1. Wet Pipe Sprinkler Systems – in this system, a positive pressure is maintained within the sprinkler system piping so that the water is immediately discharged on to the fire upon activation of the system.

2. Dry Pipe Sprinkler Systems – in this system, the sprinkler pipes are filled with pressurised air or nitrogen instead of water. The pressure in the pipe controls a remotely located 'dry pipe' valve in a closed position. As fire activates the system and the air or nitrogen is released, the dry pipe valve opens allowing the water to enter the system. The reason for use of air or nitrogen is to prevent damage to the 'exposed' areas of the system in sub-zero temperatures where water may be frozen in the pipes causing the system to be useless.

3. Deluge Systems – these are very common and effective fixed fire fighting systems on-board offshore installations. The system consists of open nozzles or sprinklers connected to the water supply system. Water is pumped into the system on demand i.e. when the deluge system is activated due to a fire alarm activated through heat, smoke or flame detection described earlier. The name 'deluge' comes from the use of a deluge valve fitted in the water supply part of the system. This valve can be operated through a hydraulic, pneumatic, electric or manual release system but when tripped, it will allow the flow of pressurised water through the system without the option to be reset.

   Deluge systems can be used in various areas such as:

   - process areas to provide protection for pipework and equipment in the area;
   - equipment requiring specific protection such as wellheads;
   - structures requiring dedicated protection;
   - areas requiring control of spread of smoke to provided safe escape routes.

Figure 6.9 A helideck team member ready to use foam monitor

4. Pre-action Systems – used in areas where high value items need protection from the effect of water if released accidentally. In operation, the system is similar to the deluge system except that the sprinkler heads or nozzles are closed. The flow of water is controlled through a valve referred to as the 'pre-action' valve due to its function within the system. The sprinkler head contains an element that operates when the space being protected reaches a certain predetermined temperature. Upon activation, the pre-action valve will open and allow the flow of water into the system but the 'sprinkling' will not commence until the sprinkler head opens.

5. Water Spray Systems – a subtype of the deluge system with the difference being that the discharge nozzle is designed to produce a spray water in a required three-dimensional shape based on the requirements within a space.

6. Water Mist Systems – also a subtype of the deluge system and used in areas where the use of water is to generate heat absorbing vapours to avoid water damage by using a lesser quantity of water. The droplet size for mist systems must be less than 1000 microns. Due to the larger surface area created by water droplets in the mist, the temperature of the space in question can be reduced quickly and more effectively.

7. Foam Water Sprinkler Systems – a mixture of foam and water is discharged through the sprinklers in this system. The piping for the system is connected to a source of foam concentrate and water supply. The fire detection system operates the flow of water through the system which is connected to the foam concentrate supply as well. This system is considered very useful for areas that contain flammable liquids. Other components of this system operate on the same principle as a deluge system.

6.9.2 Foam Systems

Foam is formed by mixing chemicals, water and air. When sprayed over any surface, it creates a blanket on the surface of burning materials particularly liquids that prevent the 'fuel' vapours from coming into contact with a source of ignition and air and therefore smothering the fire. There are two types of foam:

1. Chemical Foam – formed by mixing sodium bicarbonate (alkali) and aluminium sulphate (acid) with water. A stabiliser may be added to increase the strength of foam bubbles. When mixed, these components generate foam where bubbles are filled with CO_2. The chemicals required to produce this type of foam are available in pre-mixed ready-to-use containers that can be introduced to a fire fighting system using a device known as 'foam hopper'.

2. Mechanical Foam – This type of foam is produced by mixing foam concentrate, water and air. Normally a 3 per cent or 6 per

cent concentrate will be used which means 3 or 6 per cent of the concentrate will be mixed with 97 or 94 parts of water to generate the mixture.

The foam making procedure requires a foam generator that mixes the foam concentrate with an appropriate quantity of water and subsequently introduces air to generate bubbles i.e. foam.

Mechanical foams can be generated from the following types of materials:

1. Protein (P) foam produced by using waste protein materials from animal horns or hoofs, vegetable waste and mineral salts.
2. Fluoroprotein (FP) foam formed by adding fluorinated surfactants to the protein foam. Fluorinated surfactants are chemical compounds that reduce the surface tension of water bubbles thus increasing their life.
3. Film Forming Fluoroprotein (FFFP) foam is a subtype of fluoroprotein foam where the level of fluorochemical surfactants is increased to make a better film on hydrocarbon fuels. As a result, they offer a better fire knockdown rate and burn-back resistance. However, these can only be used on hydrocarbon liquid fires.
4. Aqueous Film Forming Foams (AFFF) are produced by using synthetic foaming agents such as hydrocarbon surfactants, solvents, freezing point depressants, fluorochemical surfactants and mineral salts. Their action on fire is similar to FFFP foam except that they form an aqueous film on the surface of flammable liquid.
5. Alcohol Resistant – Aqueous Film Forming Foam (AR-AFFF) – certain liquids can mix with water easily as opposed to the hydrocarbon liquids. Fires involving liquids mixable with water are difficult to smother due to the water in the foam mixing with the flammable liquid and thus destroying the film formed by the foam such as AFFF or FFFP. In order to avoid this problem, polymer is added to the AFFF to produce a physical barrier between the fuel and foam layer. This barrier appears like a gel when formed on the surface of the flammable liquid.

Foams are classified on the basis of expansion ratio of foam concentrate and the produced foam as below:

1. Low expansion – ratio of less than or equal to 1:20
2. Medium expansion – ratio of more than 1:20 but less than or equal to 1:200
3. High expansion – ratio of more than 1:200.

Generally foam is produced in two stages:

• in the first stage, foam is mixed with water to produce a foam solution prior to its delivery to the foam making equipment;

- in the second stage, foam solution needs to be mixed with air to produce bubbles i.e. the foam.

On offshore installations, different types of foam delivery systems are used depending upon the location where it is required. For example, on the helideck of an offshore installation, a displacement pressure proportioner with a foam concentrate tank (also known as a bladder tank) and the required number of foam monitors are placed around the helideck.

Proportioning devices ensure that a required percentage of foam concentrate is fed into the water. These devices could be in-line eductors or specialised pumps. An important component of the delivery system is the foam nozzle which must provide proper expansion of the delivered foam. Standard fog nozzles are considered unsuitable for optimum foam delivery to the fire. Foam monitors are permanently installed 'nozzles' that can discharge a large quantity of foam to the desired location such as helidecks.

6.9.3 Total Flooding Fire Suppression

Total flooding fire suppression systems are commonly based on different types of gases as a fire extinguishing medium, hence are also referred to as 'gaseous fire suppression'. The common systems found in the offshore industry are Inergen, CO_2, FM200 and Novec 1230. The principle for gaseous fire suppression is to reduce the oxygen content in the atmosphere to a level where combustion cannot be sustained.

Inergen   This is an inert gas consisting of a mixture of CO_2 (8 per cent), argon (40 per cent) and nitrogen (52 per cent). When this mixture is discharged into a space containing fire, air is mixed into the Inergen changing to 3.2 per cent CO_2, 17 per cent argon, 67.3 per cent nitrogen and 12.5 per cent oxygen. Due to its oxygen content, Inergen is considered to be extremely useful for fire protection of populated work spaces meaning that personnel can work normally to deal with an emergency in spaces where Inergen has been released. Whilst the oxygen percentage is low, the CO_2 percentage is high meaning that with an increased level of CO_2, the rate of respiration will increase meeting the body's demand for oxygen due to less oxygen concentration. Further benefits include its ability to disperse leaving no residues after the fire has been controlled, non-corrosiveness and non-damaging to the environment. Inergen is stored in cylinders and delivered to the required spaces by dedicated pipelines.

Figure 6.10 CO_2 flooding system

Figure 6.11 Novec Total Flooding System for small area

CO$_2$

This is a colourless and odourless gas consisting of carbon and oxygen. It is considered an effective extinguishing medium as it is 50 per cent heavier than air and therefore settles on fire displacing oxygen plus it provides some cooling effect. When oxygen content is reduced to below 15 per cent, the fire can be extinguished. It is important that the required CO$_2$ content is maintained for any such duration in which the burn-back effect can be overcome by reducing the temperature of the area where CO$_2$ has been released. It can therefore be used in areas containing materials that can cause Class A, B or C fires. However, the major disadvantage in comparison with Inergen is that it does not support breathing.

Novec 1230

Also known as FluoroKetone, this is a colourless, low odour fluid which is 11 times heavier than air. It can be used on fires involving flammable liquids, gases and electrical equipment. The fluid is stored in liquid form at room temperature and smothers the fire by cooling it down. The discharge time given for this system is 30 seconds. Nitrogen gas is used to propel the fluid through the system. Novec is considered to be non-conductive, non-corrosive and therefore safe for equipment. In addition it is also considered safe for personnel when the fluid is released. The danger, however, is that Novec decomposes if exposed continuously to temperatures over 500°C producing halogen acids.

FM200

DuPont's FM200, also known as heptafluoropropane, is a heavier than air, colourless and odourless gaseous suppression agent and is considered an environmentally safe alternative for halon based extinguishing agents. Available as a liquefied compressed gas, it is stable at room temperature but like Novec, it decomposes at high temperatures. The maximum discharge time given for this system is ten seconds. Due to FM200 being a non-conductive and non-corrosive material, it can be used on Class C fires to avoid damage to the equipment in a space protected by this system. It is safe to use in populated work spaces and its removal can be carried out by simple ventilation of the space through the air-conditioning or ventilation system.

6.10 Precautions when Fighting Fires

An alarm should be raised as soon as a fire is discovered. If the fire is small, then it may be tackled by use of a portable fire extinguisher or other means described earlier in this chapter. Before making an attempt to fight a fire, the escape route should be identified to ensure a safe escape for an unsuccessful attempt to control a fire. If the fire can't be controlled, then personnel must proceed to the muster station. Further action will be advised on the basis of duties assigned according to the muster list. The acronym PASS should be used to remember the sequence in which portable fire extinguishers can be used.

Pull the pin.
Aim extinguisher nozzle at the fire (or base of fire depending upon the type of fire/extinguisher used).
Squeeze the lever to trigger the extinguisher.
Sweep the extinguisher from one side to the other to ensure all surface of material on fire is covered.

General precautions for dealing with various types of fires are given below.

1. Time taken to respond to a fire will determine the extent of damage i.e. the quicker the response, the less will be the damage.
2. A few moments spent on planning may save considerable time when fighting a fire.
3. Fire intrinsically travels upwards, especially if stairways or ducts provide a chimney-like passage.
4. Escape doors are usually designed to open in the direction of travel. However, some doors may not comply with this requirement. Any such doors should be handled carefully as described in 'negotiating doors' (described later in this chapter).
5. Some personnel may be tasked to contain the spread of fire by simple techniques such as boundary cooling or removing 'fuel' from the vicinity of the space on fire.
6. Liquid fires may originate from the large volume of flammable vapours at room temperatures. Flammable liquids are generally lighter than water and therefore float. Any actions taken to tackle a liquid fire must not cause spillage of the flammable liquid. This may easily happen from the jet of even a portable extinguisher or a hose.
7. The source of the power supply must be disconnected for fires involving electrical equipment not only to remove the source of ignition but also to avoid risk of electrocution when using an extinguishing medium that may conduct electricity.
8. Appropriate PPE must be worn when fighting fires.
9. When using a portable fire extinguisher, the user must verify the authorised usage on the type of fire being extinguished. A short

Figure 6.12 Vertical Escape Ladder

Figure 6.13 Smoke hood (Source: Ceronav Romania)

burst can be used to establish correct operation of the extinguisher. In all cases, full contents must be discharged as they will not be of any use subsequent to partial use. If one extinguisher is not sufficient, then an additional extinguisher should be used to smother the fire completely. Half-extinguished fires will not solve the problem.

10. Once a fire is extinguished, the temperature of the space should be monitored for sufficient duration (at least 30 minutes) to monitor possibility of re-ignition.

11. For gas fires, the supply of gas must be removed by closing valves before tackling the fire.

12. When using SCBA (Self Contained Breathing Apparatus), availability of the air supply should be checked by the wearer to allow sufficient time to reach the space on fire, fight the fire and return to safety.

6.10.1 Escaping from Fire

Escaping from fire can be quite critical for ensuring a safe escape. Agreed procedures must be practised in drills to put them into practice when required. A group escape should be preferred where possible over an individual escape. When in groups, assistance can be offered to others who are injured or lack the knowledge of escape routes. The general procedure for escaping from a fire and reaching a safe place such as a muster station is given below.

1. Each individual worker on an offshore installation will be provided with a smoke hood which can prove to be vital. This must be fully donned to prevent damage from smoke.

2. Remain calm, panic won't help at all.

3. When in groups, maintain physical contact with other people in the group. This can easily be achieved by, for example, walking in a line with the person in front checking the escape route and the person behind keeping one hand on the shoulder of the person in front. The second hand can be used to maintain contact with the bulkhead, railings or boundaries of the escape route.

4. Smoke and hot gases flow towards deck heads/ceilings, therefore keep as low as possible to remain as clear as possible.

5. If possible, carry a portable extinguisher or a fire hose which may prove to be useful if confronted by a fire when on the escape route.

6. Walk slowly checking the integrity of the surface being walked on as it may have been weakened due to fire in the compartments below. Fire resistant composite deck gratings have known to fail due to fires, hence extra care must be exercised in any areas where decks are made of gratings.

7. If it becomes difficult to keep standing or walking due to smoke or heat, then crawl on the deck.
8. When required to open any doors, keep in mind the precautions stated in the section below on 'negotiating doors'. Do not leave any open doors behind to prevent the spread of fire.

6.10.2 Using Smoke Hoods

Smoke hoods, also known as emergency escape hoods, are designed specifically for use in situations where an escape from smoke resulting from fire may be required. They are designed for one use only and must not be used for any other purposes. They can be donned simply by putting two hands through the base and sliding the hood over the head. They can be used with spectacles and will protect the wearer for up to 15 minutes as described below:

- head from heat since they are made of heat resistant materials;
- eyes from irritation caused by smoke or other fumes caused by fire;
- breathing from inhalation of harmful gases by use of filters that can remove CO and other gases.

Whilst the personal smoke hoods are supplied in sealed bags to be used only in emergency, training hoods can be available to use for training and familiarisation. Hoods are required to comply with certain design and approval requirements[1] such as:

- capable of being donned within 30 seconds;
- capable to remain in secure donned position;
- their function should not be affected by water deluge system;
- ability of the wearer to carry out routine movements should not be restricted;
- any valves fitted to the breathing circuit shall be easy to operate;
- visor, if fitted, shall be securely attached and capable of withstanding normal handling;
- any damage to packaging or opened seal for the packaging should be easily identifiable.

There are two main types of smoke hoods:

1. Filter Smoke Hoods – fitted and tested for use with Carbon Monoxide/Dioxide, Sulphur Dioxide, Hydrogen Chloride and Hydrogen Sulphide gases.
2. Breathable Gas Smoke hoods – these hoods are fitted with a source of compressed breathable gas (air or oxygen). These hoods are tested to provide up to 15 minutes of breathable gas with 10 minutes allowed for low level exercises and 5 minutes of high level exercise with carbon dioxide inhalation

allowance not more than 1 per cent and oxygen inhalation level not below 19.5 per cent. These hoods should also allow the wearers to communicate with each other at a distance of 2 metres.

6.10.3 Negotiating Doors and Stairways

Procedures for opening/closing doors to the spaces on fire must be practised during training and drills. These exercises must include the use of appropriate PPE and RPE in addition to the equipment used for fire fighting. When negotiating doors, the following precautions must be adhered to:

1. Before attempting to open a door with a known fire inside the compartment, a clear escape route should be established in case an escape or evacuation is required.
2. Suitable water supply e.g. from a live fire hose or portable extinguisher must be available and ready to use.
3. Communication between the members of the team and the control station must be tested prior to commencing the task and established throughout the task.
4. The temperature of the door can be checked by touching it with the back of the hand without gloves. Extreme caution should be exercised when checking any door with the hand since a hot surface may burn the skin. The temperature of the door should be checked at various heights, for example, at the top of the door, middle and bottom to ascertain the position of the fire in the compartment.
5. Boundary cooling can help reduce the temperature of the surroundings of the compartment to provide a workable temperature.
6. When opening the door, the person taking the lead must use the door as a shield from smoke or flame. Knees could be used to provide some pressure on the door whilst sitting on the deck to open it by turning the door handle. Remember, the handle could be hotter than the remaining parts of the door, especially if the door is made of non-metal material with a metal handle.
7. When set in position, open the door slowly to let the smoke and heat escape above the head level. This will also allow an opportunity for a second person in the fire fighting team to assess the extent of the fire within the space. Once assessed and if it is safe to do so, the door can be opened fully and fire tackled by using appropriate fire fighting means. This procedure can also be used for escaping through alleyways, etc. when the adjacent compartments may be on fire.
8. When personnel have to use stairways to escape to a muster station or a safe place, the above general precautions should be considered but in addition:

Figure 6.14 Negotiating doors
(Source: Ceronav Romania)

a. one person should be on any one tread of the stairs at any one time;
b. integrity of each tread should be tested by putting the load of one foot first before placing the second foot on the same tread;
c. railings should be held on to at all times;
d. always face the stairs whether climbing up or down.

Note

1 HSE (1997) Smoke Hoods for Offshore Use – Performance Requirements. Offshore Technology Report OTO 96 703. Available at www.hse.gov.uk/research/otopdf/1996/oto96703.pdf [Accessed 17.11.14].

7

Survival at Sea

7.1 Offshore Emergencies

Offshore emergencies can be highly complex and therefore all personnel involved in dealing with them must be fully trained and prepared to use all equipment and assets at their disposal. Typical foreseeable emergencies that may be encountered on an offshore installation include:

1. fire or explosion that may be combined with any of the emergencies given below;
2. structural failure due to fatigue or corrosion;
3. stability of the structure;
4. extreme weather;
5. other vessels colliding with the structure;
6. failure of mooring systems causing the installation to move from its position;
7. accidents to the helicopters or support vessels;
8. blowouts or uncontrolled flow from a well;
9. H_2S gas incidents;
10. fall from height;
11. accidents related to enclosed spaces such as during tank entry for cleaning;
12. oil or other hydrocarbon spills.

In the UK, the PFEER regulations require duty holders to identify circumstances that may necessitate an evacuation, escape or rescue from an installation. Offshore Installations (Management and Administration) Regulations 1995 require the duty holder to put in place a system that enables installation to monitor POB (persons on-board) so that an appropriate head count can be taken in the event of an emergency. Whilst each emergency can be considered separately for the response required to ensure safety of life, environment and property mostly, one problem (such as fire) may have resulted from another problem such as hydrocarbon release. As a consequence, all personnel must keep in mind when preparing for emergencies that the personnel may have been prepared for 'generic' standard scenarios but the actual emergency may be completely different. Hence, the better the response, the easier it will be to tackle the emergency.

The end result required for an offshore installation is continuity of the operations for which it is deployed. Therefore an important aspect to consider during risk assessment and mitigation of consequential emergencies must incorporate suitable means to resume those operations safely and as soon as possible. All major installations will devise an Emergency Response Strategy (ERS) based on findings of the risk assessments. ERS will define the exact response required for each hazard identified in the risk assessment and resources required to mitigate the risks. For example, the resources at the disposal of OIM may include personnel, equipment, support vessels and helicopters. Additional resources may have been identified from the surrounding areas such as support vessels and helicopters from other installations in the vicinity. In addition, assistance available from national or international authorities such as SAR services, coastguards, oil spill response units, health services or even defence forces may need to be identified and procedures established to call upon these resources. Upon establishing availability of the internal or external resources, an Emergency Response Plan (ERP) will be established to stipulate exact responsibilities and duties for each identified resource as well as the provision of equipment to enable escape, evacuation, rescue or recovery of personnel following an emergency. As per the PFEER regulations, an ERP is defined as 'a document setting out the organisation, arrangements and procedures for use in an emergency'. The most important aspect of any ERP is to define a 'chain of command' so that effective decision making is not compromised at the time of acute need.

7.2 Emergency Signals and Procedures

In line with an installation's ERS and ERP, procedures will be put into practice as soon as an emergency strikes. This section provides generic information about these procedures and the general set-up found on offshore installations. Each installation will have specific procedures to be followed to deal with emergencies on-board that installation. These procedures may vary significantly for another installation due to differences in the area of operation, nature of tasks, competencies of personnel or for other reasons. This is why all new joiners will receive a general induction upon arrival on the installation. This covers all aspects of their stay on-board but specific guidelines are provided for actions to be taken in case of an emergency. Reminder 'notices' written in English or the most common working language on-board the installation are placed in workers' cabins to show:

- location of allocated muster station;
- duties in specific emergencies;

Transocean — DEEPWATER HORIZON STATION BILL

CHAIN OF COMMAND

MASTER
CHIEF OFFICER
OFFSHORE INSTALLATION MANAGER

Master		O.I.M.
Master		O.I.M.

EMERGENCY CONTACT

To raise the alarm in the event of any emergency, contact the CCR/Bridge on phone #124

EMERGENCY SIGNALS

Fire & Emergency	Seven (7) or more short blasts followed by one (1) long blast on the unit's General Alarm supplemented by the unit's whistle for a period of not less than 10 seconds.
Prepare to Abandon	Continuous sounding of the general alarm supplemented by the unit's whistle for a period not less than 10 seconds.
	The order to Abandon will be given verbally by the Master.
Man Overboard	Hail, and pass the words Man Overboard and 3 long soundings of the unit's whistle to be repeated 3 times.
LEL Gas (High Level)	BLUE flashing light and a High-Low, Two-Tone Alarm
H2S Gas (Low Level – 5 PPM)	AMBER flashing light
H2S Gas (High Level – 10 PPM)	AMBER flashing light, Yelping siren / warbling horn.
All Clear	Three short soundings of the general alarm and/or unit's whistle

These signals shall be accompanied by an announcement on the rig's PA system.

GENERAL INSTRUCTIONS

1. All personnel shall familiarize themselves with the location and duties of their 'Fire & Emergency Stations', 'Lifeboat Stations', 'Alternate lifeboat Stations' and 'Emergency Signals' as shown on the Station Bill immediately upon reporting onboard.
2. All personnel shall ensure that there is a serviceable lifejacket, smoke hood, lightstick and gloves for each bunk in his/her cabin. Notify the CCR/Bridge immediately if there are not enough lifejackets in the cabin. Spare lifejackets are located at the lifeboat muster stations for emergency teams and tour personnel and visitors.
3. All personnel shall attend an orientation briefing on board the vessel. The briefing will include instructions in the contents of this Station Bill and other Rig specific instructions.
4. All personnel will participate in all 'Emergency Drills' as if it were an actual emergency. All personnel will be dressed in full work attire including general PPE. The Master is the only person with the authority to excuse personnel from attending emergency drills.
5. During periods of rough weather or hazardous operations all watertight doors and openings are to be kept closed. All watertight doors in the columns to be kept closed at all times except for normal traffic.
6. Any person discovering a fire shall immediately do the following:
   a. Raise the alarm (activate manual call point, call emergency contact number, or utilize any other means of raising the alarm.)
   b. Contain or fight the fire using available equipment (Without risking his or her own safety).
   c. Evacuate the area if hazardous to personal safety.
7. On detection of any Oil/Chemical or Liquid Mud spill, the CCR/Bridge must be informed immediately and measures started to contain the spill utilizing available spill control equipment.
8. All accidents, incidents, and/or hazardous conditions must be reported immediately to Supervisor and CCR/Bridge.
9. While onboard all Operator, Visitor and Service Personnel shall follow all Transocean safety instructions, rules and regulations. Failure to do so will result in their removal from the rig.
10. Responsibility for GMDSS communications will be assigned by Master.
11. Chief Mate is responsible for the inspection and maintenance of emergency and lifesaving equipment.

Note: See SOLAS training manual for additional information (IMO MODU Code 14.8.16)

ABANDONMENT STATIONS

LIFEBOAT #1
ALTERNATIVE : LIFEBOAT #3
LIFERAFT: FORWARD

MASTER	IN COMMAND
DPO (Off)	2nd I/C. COXWAIN
Chief Mechanic (Engine) - Off	3rd I/C.
A/B SEAMEN	PREPARE LIFERAFTS
	Assist as Directed
Asst. DRILLER (Off-Rm 333)	TAKE MUSTER

2nd I/C is responsible for VHF/SART/EPIRB etc
ALL OTHER PERSONS
IN ROOM NUMBERS

227, 228, 229, 231, 233, 237, 239, 241, 243, 249, 303, 305, 307, 309, 311, 313, 317, 319, 329, 331, 333, 335, 337, 339, 341, 343, 345, 347, 349, 351, 353, 355, 357	AS DIRECTED BY COXSWAIN, ENTER BOAT AND ANSWER MUSTER

LIFEBOAT #2
ALTERNATIVE : LIFEBOAT #4
LIFERAFT: AFT

CHIEF MATE	IN COMMAND
Sr. DPO (Off)	2nd I/C. COXWAIN
BOSUN	3rd I/C.
A/B SEAMAN	PREPARE LIFERAFTS
	Assist as Directed
Asst. DRILLER (Off-Rm 334)	TAKE MUSTER

2nd I/C is responsible for VHF/SART/EPIRB etc
ALL OTHER PERSONS
IN ROOM NUMBERS

226, 230, 232, 234, 236, 238, 240, 244, 248, 302, 304, 308, 310, 312, 314, 316, 318, 320, 322, 328, 330, 332, 334, 336, 338, 340, 342, 344, 346, 348, 352, 354, 356	AS DIRECTED BY COXSWAIN, ENTER BOAT AND ANSWER MUSTER

Helideck

FWD

Accommodations

Lifeboat 2 Rescue Boat
Lifeboat 1

Lifeboat #2 Muster Area

Lifeboat #1 Muster Area

Fire Team #1 Muster Area

PORT

STBD

Lifeboat #4 Muster Area

Lifeboat #3 Muster Area

Fire Team #2 Muster Area

Lifeboat 4

Lifeboat 3

AFT

● Emergency Escape to Water
▭ Liferaft

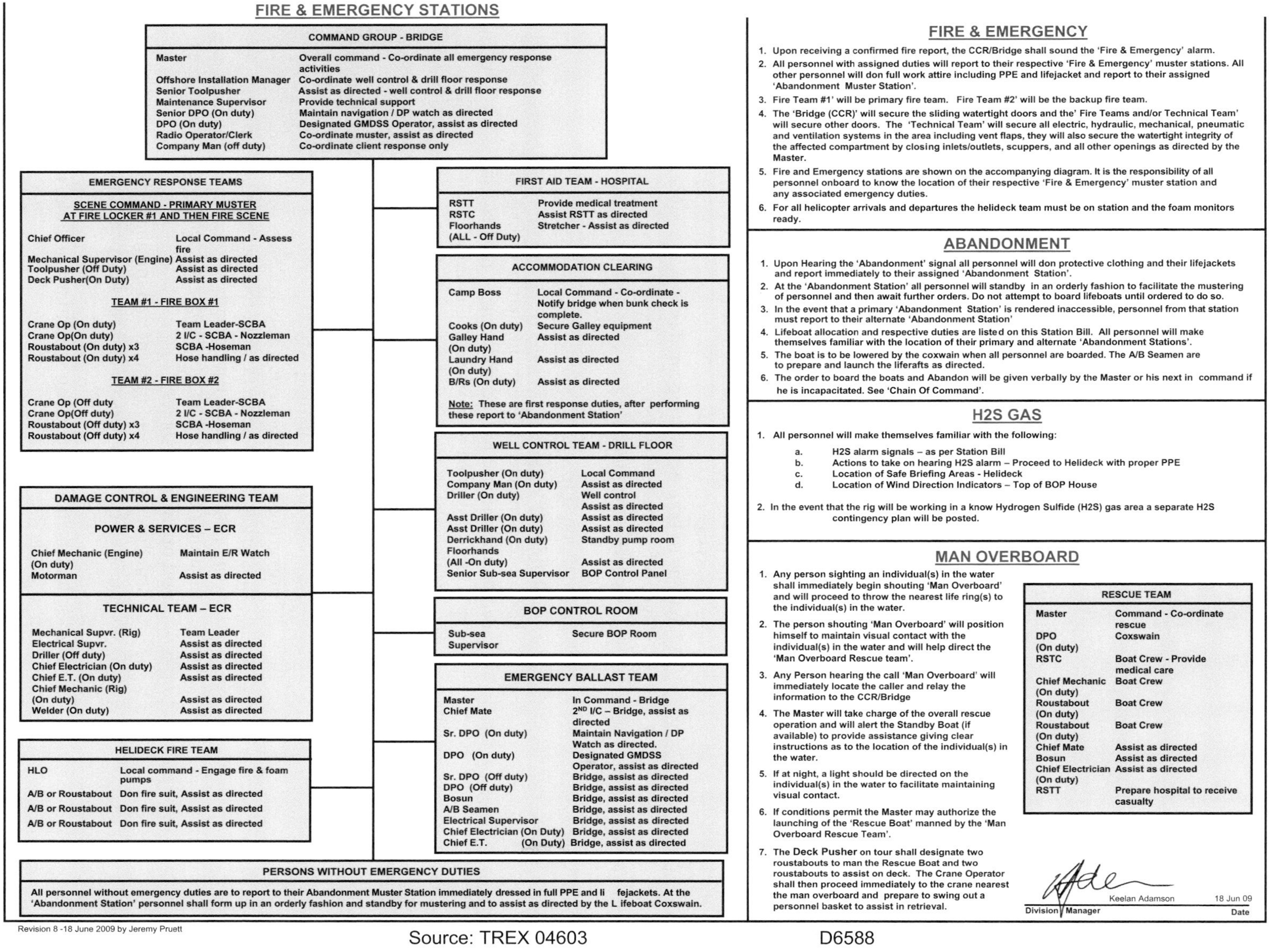

FIRE & EMERGENCY STATIONS

FIRE & EMERGENCY

1. Upon receiving a confirmed fire report, the CCR/Bridge shall sound the 'Fire & Emergency' alarm.
2. All personnel with assigned duties will report to their respective 'Fire & Emergency' muster stations. All other personnel will don full work attire including PPE and lifejacket and report to their assigned 'Abandonment Muster Station'.
3. Fire Team #1 will be primary fire team.   Fire Team #2 will be the backup fire team.
4. The 'Bridge (CCR)' will secure the sliding watertight doors and the 'Fire Teams and/or Technical Team' will secure the other doors. The 'Technical Team' will secure all electric, hydraulic, mechanical, pneumatic and ventilation systems in the area including vent flaps, they will also secure the watertight integrity of the affected compartment by closing inlets/outlets, scuppers, and all other openings as directed by the Master.
5. Fire and Emergency stations are shown on the accompanying diagram. It is the responsibility of all personnel onboard to know the location of their respective 'Fire & Emergency' muster station and any associated emergency duties.
6. For all helicopter arrivals and departures the helideck team must be on station and the foam monitors ready.

ABANDONMENT

1. Upon Hearing the 'Abandonment' signal all personnel will don protective clothing and their lifejackets and report immediately to their assigned 'Abandonment Station'.
2. At the 'Abandonment Station' all personnel will standby in an orderly fashion to facilitate the mustering of personnel and then await further orders. Do not attempt to board lifeboats until ordered to do so.
3. In the event that a primary 'Abandonment Station' is rendered inaccessible, personnel from that station must report to their alternate 'Abandonment Station'.
4. Lifeboat allocation and respective duties are listed on this Station Bill. All personnel will make themselves familiar with the location of their primary and alternate 'Abandonment Stations'.
5. The boat is to be lowered by the coxwain when all personnel are boarded. The A/B Seamen are to prepare and launch the liferafts as directed.
6. The order to board the boats and Abandon will be given verbally by the Master or his next in command if he is incapacitated. See 'Chain Of Command'.

H2S GAS

1. All personnel will make themselves familiar with the following:

   a.   H2S alarm signals – as per Station Bill
   b.   Actions to take on hearing H2S alarm – Proceed to Helideck with proper PPE
   c.   Location of Safe Briefing Areas - Helideck
   d.   Location of Wind Direction Indicators – Top of BOP House

2. In the event that the rig will be working in a know Hydrogen Sulfide (H2S) gas area a separate H2S contingency plan will be posted.

MAN OVERBOARD

1. Any person sighting an individual(s) in the water shall immediately begin shouting 'Man Overboard' and will proceed to throw the nearest life ring(s) to the individual(s) in the water.
2. The person shouting 'Man Overboard' will position himself to maintain visual contact with the individual(s) in the water and will help direct the 'Man Overboard Rescue team'.
3. Any Person hearing the call 'Man Overboard' will immediately locate the caller and relay the information to the CCR/Bridge
4. The Master will take charge of the overall rescue operation and will alert the Standby Boat (if available) to provide assistance giving clear instructions as to the location of the individual(s) in the water.
5. If at night, a light should be directed on the individual(s) in the water to facilitate maintaining visual contact.
6. If conditions permit the Master may authorize the launching of the 'Rescue Boat' manned by the Man Overboard Rescue Team'.
7. The Deck Pusher on tour shall designate two roustabouts to man the Rescue Boat and two roustabouts to assist on deck. The Crane Operator shall then proceed immediately to the crane nearest the man overboard and prepare to swing out a personnel basket to assist in retrieval.

COMMAND GROUP - BRIDGE

Master	Overall command - Co-ordinate all emergency response activities
Offshore Installation Manager	Co-ordinate well control & drill floor activities
Senior Toolpusher	Assist as directed - well control & drill floor response
Maintenance Supervisor	Provide technical support
Senior DPO (On duty)	Maintain navigation / DP Watch as directed
DPO (On duty)	Designated GMDSS Operator, assist as directed
Radio Operator/Clerk	Co-ordinate muster, assist as directed
Company Man (off duty)	Co-ordinate client response only

EMERGENCY RESPONSE TEAMS

SCENE COMMAND - PRIMARY MUSTER AT FIRE LOCKER #1 AND THEN FIRE SCENE

Chief Officer	Local Command - Assess fire
Mechanical Supervisor (Engine)	Assist as directed
Toolpusher (Off duty)	Assist as directed
Deck Pusher(On duty)	Assist as directed

TEAM #1 - FIRE BOX #1

Crane Op (On duty)	Team Leader-SCBA
Crane Op(On duty)	2 I/C - SCBA - Nozzleman
Roustabout (On duty) x3	SCBA -Hoseman
Roustabout (On duty) x4	Hose handling / as directed

TEAM #2 - FIRE BOX #2

Crane Op (Off duty)	Team Leader-SCBA
Crane Op(Off duty)	2 I/C - SCBA - Nozzleman
Roustabout (Off duty) x3	SCBA -Hoseman
Roustabout (Off duty) x4	Hose handling / as directed

DAMAGE CONTROL & ENGINEERING TEAM

POWER & SERVICES - ECR

Chief Mechanic (Engine) (On duty)	Maintain E/R Watch
Motorman	Assist as directed

TECHNICAL TEAM - ECR

Mechanical Supvr. (Rig)	Team Leader
Electrical Supvr.	Assist as directed
Driller (Off duty)	Assist as directed
Chief Electrician (On duty)	Assist as directed
Chief E.T. (On duty)	Assist as directed
Chief Mechanic (Rig) (On duty)	Assist as directed
Welder (On duty)	Assist as directed

HELIDECK FIRE TEAM

HLO	Local command - Engage fire & foam pumps
A/B or Roustabout	Don fire suit, Assist as directed
A/B or Roustabout	Don fire suit, Assist as directed
A/B or Roustabout	Don fire suit, Assist as directed

PERSONS WITHOUT EMERGENCY DUTIES

All personnel without emergency duties are to report to their Abandonment Muster Station immediately dressed in full PPE and lifejackets. At the 'Abandonment Station' personnel shall form up in an orderly fashion and standby for mustering and to assist as directed by the Lifeboat Coxswain.

FIRST AID TEAM - HOSPITAL

RSTT	Provide medical treatment
RSTC	Assist RSTT as directed
Floorhands (ALL - Off duty)	Stretcher - Assist as directed

ACCOMMODATION CLEARING

Camp Boss	Local Command - Co-ordinate - Notify bridge when bunk check is complete.
Cooks (On duty)	Secure Galley equipment
Galley Hand (On duty)	Assist as directed
Laundry Hand (On duty)	Assist as directed
B/Rs (On duty)	Assist as directed

Note: These are first response duties, after performing these report to 'Abandonment Station'

WELL CONTROL TEAM - DRILL FLOOR

Toolpusher (On duty)	Local Command
Company Man (On duty)	Assist as directed
Driller (On duty)	Well control
Asst Driller (On duty)	Assist as directed
Asst Driller (On duty)	Assist as directed
Derrickhand (On duty)	Standby pump room
Floorhands (All -On duty)	Assist as directed
Senior Sub-sea Supervisor	BOP Control Panel

BOP CONTROL ROOM

Sub-sea Supervisor	Secure BOP Room

EMERGENCY BALLAST TEAM

Master	In Command - Bridge
Chief Mate	2ND I/C – Bridge, assist as directed
Sr. DPO (On duty)	Maintain Navigation / DP Watch as directed.
DPO (On duty)	Designated GMDSS Operator, assist as directed
Sr. DPO (Off duty)	Bridge, assist as directed
DPO (Off duty)	Bridge, assist as directed
Bosun	Bridge, assist as directed
A/B Seamen	Bridge, assist as directed
Electrical Supervisor	Bridge, assist as directed
Chief Electrician (On Duty)	Bridge, assist as directed
Chief E.T. (On Duty)	Bridge, assist as directed

RESCUE TEAM

Master	Command - Co-ordinate rescue
DPO (On duty)	Coxswain
RSTC	Boat Crew - Provide medical care
Chief Mechanic (On duty)	Boat Crew
Roustabout (On duty)	Boat Crew
Roustabout (On duty)	Boat Crew
Chief Mate	Assist as directed
Bosun	Assist as directed
Chief Electrician (On duty)	Assist as directed
RSTT	Prepare hospital to receive casualty

Keelan Adamson   18 Jun 09
Division / Manager   Date

D6588

Source: TREX 04603

Source: www.mdl2179trialdocs.com/releases/release20130404120022/D-6588.pdf)

Figure 7.1 Deepwater Horizon station bill (Source: www.mdl2179trialdocs.com/releases/release20130404120022/D-6588.pdf)

- identification of allocated survival craft (lifeboat/liferaft);
- specific signals for:
  - general emergency;
  - abandoning the installation;
  - other emergencies.

7.2.1 Muster List

The muster list (also known as the station bill) contains information about specific duties in emergencies for workers on-board an installation. With change in personnel, their capabilities, addition of new equipment or change in procedures, changes are immediately applied to the muster list to keep it up to date at all times. Copies of the muster list are posted in conspicuous places throughout the installation including in crew accommodation spaces. Generally, the following duties are shown on the muster list in English language and, in addition, another language if required due to workers speaking a different language.

1. Instructions for operation of the following:
   a. emergency alarm signal;
   b. PA system.
2. Actions to take on hearing an alarm.
3. Signals for the following:
   a. general emergency;
   b. fire;
   c. abandon ship;
   d. man overboard.
4. Person responsible for the maintenance of LSA/FFA.
5. Person responsible for the updating of muster list.
6. The duties assigned to workers including:
   a. closing the watertight doors, fire doors, valves, scuppers, side scuttles, skylights, portholes, and other similar openings;
   b. equipping the survival craft and other life-saving appliances;
   c. preparing and launching the survival craft or other means of escape from installation;
   d. preparing other life-saving appliances;
   e. mustering the passengers and other persons on-board;
   f. using communication equipment;
   g. manning the emergency squad assigned to deal with fires and other emergencies;
   h. using fire fighting equipment and installations.
7. Nominated back-up personnel if the key personnel are disabled.

Duties during an emergency on a standard offshore installation are given in Table 7.1.

Table 7.1 – Duties in emergency	
OIM	In command on the installation. This role may be designated to a nominated emergency manager. All actions related to an emergency are carried out under direct orders of the OIM or the emergency manager. Completion of all tasks must be reported back to get authorisation for the next actions.
Deputy OIM	Directs Emergency Response Team (ERT) and the medic as required by OIM
Control Room Operator	Makes announcements on the PA as directed by the OIM and monitors equipment and other tasks as required to deal with the emergency
Muster Checker(s)	Takes a head count on muster station and reports back to control room for any missing persons and or casualties
Central Muster Checker	Nominated only on installations where large number of personnel are employed. Acts as an intermediary between the control room and muster checker(s)
Event Logger	Ensures actions are carried out according to the ERP checklists and maintains a log of events for record keeping
Radio Operator	Takes charge of communications between the installation, support vessel, shore authorities such as coastguard, health services and helicopters

For dealing with helicopter emergencies, the PFEER regulations require the following equipment placed close to the helideck in a suitable locker:

- an aircraft-type axe
- a large axe
- a heavy-duty hacksaw, with spare blades
- a grab hook
- a quick release knife
- a crowbar
- a pair of large bolt croppers.

7.2.2 Muster Stations

Muster stations are specified locations on all installations where personnel are required to report to the person in charge of that station. In cases where an emergency renders a muster station inaccessible, an alternative muster station must be designated as per the ERP. Generally, muster stations are located close to the lifeboat/liferaft embarkation areas or close to other means of escape. These areas should be:

- large enough to accommodate all personnel;
- readily accessible from accommodation and work spaces;
- fitted with emergency lighting throughout the installation.

Suitable labels should be used to give directions to the muster station.

Before proceeding to the muster station, personnel must pay attention to the announcement on the PA system to ensure they arrive at the correct muster station.

Workers on an installation may be alerted by situations that may cause threat to life, environment or property. The five human senses i.e. seeing, hearing, touching, smelling or tasting give a perception of the outside world. However, not every sense provides a similar level of interaction for information gathering. Research has shown the following levels of sense usage for gathering information: sight 80 per cent, hearing 14 per cent, touch 2 per cent, smell 2 per cent and taste 2 per cent.

Various emergency situations will have different typical characteristics noticeable by human senses. For example, an explosion will stimulate hearing before other senses come into play. The smell of burning from a fire could be the first noticeable sign before seeing the flame or feeling the rise in temperature. Regardless of how we are alerted to an emergency, the important factor to consider is the procedure through which an individual will deal with any 'notification' of emergency picked up by human senses.

The first reaction to any emergency needs to be alerting someone responsible for dealing with the emergency. This will usually be the control room needing the minimum information below to initiate an effective action:

1. Who – information about you?
2. Where – the location of emergency?
3. What – the nature of emergency?
4. When – estimated time when you spotted the emergency?

Once the control room has been alerted through a brief and quick message, you must ensure your own safety before any considerations to tackle the emergency. This may involve informing others around the area and limiting the further spread of effects of the threat. The following rule is an example for a potential fire situation that could assist in reducing the negative impact when you find yourself exposed to a threatening situation:

- Remove all personnel who may be in an immediate danger.
- Ensure spread of fire and smoke is restricted by closing door(s).
- Activate the alarm using nearest alarm activation switch.
- Call the control room and give information Who, Where, What, When as described earlier.
- Try to extinguish fire ensuring your own safety first.

As soon as a worker becomes aware of an emergency, the control room must be informed by the quickest means of communication available. These include shouting loudly the nature of emergency (e.g. Fire, Fire, Fire ...), portable radios, the installation's telephone system or activation of the installation's alarm system through alarm activation or call points. After raising the alarm initially, the control room should be contacted by alternative means to give the full details of the emergency so that a proper response can be initiated. The person becoming aware of the emergency may try to deal with the emergency but only after ensuring their own safety and if they can handle the emergency by themselves. If this is not possible, then all personnel must proceed to the muster stations to enable an appropriate response as required by the ERP procedures.

7.3 Emergency Alarms and Signals

Offshore installations are fitted with monitoring systems that monitor specified areas for potential problems such as fire or flooding, etc. In cases where pre-defined conditions are breached, automatic alarms are activated notifying workers of the problem. Whilst sensors for fire, heat, and smoke or flame detection may be fitted in various locations within the installation, the alarm monitoring panels can be located in the control room from where the exact location for which alarm is raised can be identified.

In addition to automatic sensors, installations are fitted with manual alarm activation points or break glass alarms from where the operator can break the glass to activate the alarm. It is important to note that these activation points will switch the alarm 'on' but the operator will not be able to turn them off. Once the situation has been resolved, then the alarm can be deactivated from the control room.

Offshore installations use different alarms and signals to alert personnel in various emergencies. There may be a slight variation in the signals used, hence all personnel must check their muster lists to familiarise themselves with the signals. All permits to work are 'automatically' cancelled upon activation of any alarm.

7.3.1 General Platform Alarm (GPA) or General Alarm (GA)

This alarm consists of a distinct and intermittent tone signal accompanied by a yellow flashing light through the installation. This signal is usually followed by an announcement on the PA system to inform about the nature of the emergency and the actions required. Upon hearing this alarm, all personnel must report to their muster stations unless directed otherwise. For example, the offshore medic assisted by other nominated personnel may be asked to proceed to the scene of an accident if any first aid treatment is required for injured personnel. Typical actions in the event of a general emergency include:

1. Stop all work ensuring equipment and machinery is secured in a safe manner as per the agreed procedures or as directed by the OIM to avoid any other mishap.
2. All personnel must don appropriate warm clothing, lifejacket and survival suit as required by the nature of emergency prior to arriving at the muster station. Most offshore installations provide grab bags containing the necessary equipment/PPE such as flash light, fire resistant gloves, smoke hood for abandoning the installation.
3. Workers with duties under the ERP to respond to emergencies (collectively referred to as essential personnel) such as fire or damage must be prepared to deal with them as per the ERP.
4. Visitors, contractors or other personnel (collectively known as non-essential personnel) who do not have any ERP assigned duties must report fully prepared with their lifejackets, survival suits and warm clothing to the allocated muster station.
5. Once at the muster station, a roll call will be taken by the designated 'muster checker' and personnel briefed about the nature of emergency and further actions required. T-Cards or a swipe card system may be used to take a roll call or head count. Depending upon the total number of personnel on-board an installation, there may be more than one muster station on-board. Each muster station will be looked after by a muster checker who will report to the 'central muster coordinator'. When the central muster coordinator is satisfied with the total head count on the muster station being equal to the total POB, a response is initiated. However, a search may be organised through search parties if any personnel are missing from the muster station.

7.3.2 Signal to Abandon Platform/Installation

When it becomes clear to the OIM that the only option to ensure safety of life is by abandoning the installation, then a Prepare to Abandon Alarm (PAPA) will be sounded. Generally this alarm is continuous ringing of the siren or alarm which requires all personnel to proceed to abandon the platform muster station.

An order to abandon platform is given through verbal command from the OIM or the most senior surviving officer and may be transmitted on the platform's public address system. Upon hearing this order, all personnel should embark the survival craft which will be launched by designated personnel. Other means such as helicopter, lifeboats, liferafts, etc. may also be used to abandon the installation.

7.3.3 Man Overboard (MOB) Signal

MOB is indicated by shouting 'man overboard' loudly which will be followed by an announcement over the public address system. The person seeing someone falling overboard should immediately throw the closest lifebuoy to assist the MOB and give full details to the control room by using suitable means of communication. The MOB should never be allowed to disappear from sight as relocation will be very difficult, especially in extremely rough sea conditions causing heavy swell. Once the control room is aware of the situation, they can direct the ERRV which will take necessary actions to recover the MOB by using recovery devices or FRC (Fast Rescue Craft). For installations where no ERRV is available, designated personnel will be required to prepare and launch their own FRC to recover the MOB.

7.3.4 Escape Routes and Safety Signs

Installations are required to be provided with either direct escape routes to the evacuation point or from working/accommodation areas to temporary refuges and then to the evacuation point. These routes and refuges are required to comply with minimum endurance times to allow sufficient time for reaching the evacuation point. This is ensured by adding layers of safety through enhanced structure for the areas in question as described in Chapter 6.

Access and egress areas are identified and marked with suitable luminescent signs/placards with emergency lighting also provided as back-up for failure of the main power supply. All personnel must familiarise themselves with these signs, layout of the place, exit doors, escape routes, call points and nearest telephones. It will prove to be extremely useful to know the secondary means of escape and communication in case the primary means have been compromised.

Escape signs are green in colour printed with white labels or pictograms. These signs are required to either be illuminated by using emergency lighting, luminescent or both. PFEER regulations require the following areas to be fitted with adequate emergency lighting and suitable signs to ensure safe access to the muster station:

Figure 7.2 Fast rescue craft

Figure 7.3 Various combinations of emergency exit signs

- exits from accommodation and work spaces;
- access routes to muster stations;
- access routes to temporary refuges;
- access routes to evacuation and escape points;
- evacuation and escape areas.

Emergency lighting is considered to be the Safety Critical Element (SCE) and therefore special requirements are needed for testing and functionality. The areas that require specific consideration for provision of emergency lighting include:

- lifeboat embarkation areas;
- helideck and access route thereto;
- fire pump and sprinkler control area;
- emergency response equipment lockers;
- radio/communications room, cargo control room, etc.;
- muster stations.

7.3.5 Training and Drills

Training and drills help practice procedures and use of equipment that may be required in emergencies. Keeping in mind that no two emergencies will have the same sequence of events and therefore consequences, it is down to the personnel dealing with those emergencies to improve the success rate.

In any emergency, even well-practised individuals are likely to suffer from various levels of anxiety and stress caused by panic when there is:

- a lack of information about the situation combined with a lack of direction;
- a perception of grave and imminent danger to life;
- a feeling of helplessness due to a feeling of being trapped.

It has been proven scientifically that the human body releases a hormone known as 'cortisol' during stress. This hormone is known to cause memory loss. Consequently, the memorised 'emergency' procedures can be forgotten in a real life emergency. However, when individuals are involved in drills, they practise procedures and use of equipment and therefore train their reflexes to follow the prompt. Consequently, the 'memory' may not play a vital role in the actions they are required to take and hence their 'reactions' will correspond to the procedures. This will result in improving the response to any emergency by overcoming the human weakness caused by the impulsive behaviours leading to stress.

Preparation to Abandon Platform drill consists of a muster of the workers (and of visitors/contractors, if appropriate) at the stations referred to in the muster list, and a muster and drill at survival craft stations. It is recommended that a fire drill be held simultaneously with the first stage of the abandon ship drill. Drills for emergencies other than fire, e.g. collision, damage control, or spillage, rescue of personnel from dangerous spaces, or medical treatment, may be conducted in lieu of or in addition to a fire drill, provided every crew member participates in at least one fire drill each month. Whether a fire or other emergency drill is to be conducted, it is useful to discuss beforehand the objectives of the drill in order that those taking part can derive the maximum benefit from the drill.

Platform abandon drills must be carried out in a manner simulating a real emergency. TEMPSC or other lifeboat engines must be operated for a period of not less than five minutes. Further details of procedures related to TEMPSC are given later in this chapter.

Each platform abandon drill shall include:

1. summoning of crew and all other personnel on-board to muster stations with the ship's emergency alarm;
2. reporting to muster stations and preparing for the duties described in the muster list;
3. checking that crew and all other personnel are suitably dressed;
4. checking that lifejackets are donned;
5. lowering of at least one lifeboat after any necessary preparation for launching;
6. starting and operation of the lifeboat engine;
7. operation of the davits used for launching liferafts;
8. a mock search and rescue of personnel trapped in their cabins; and
9. instructions in the use of radio and other life-saving appliances.

For each drill, records should be maintained for the activities carried out and the personnel who attended the drill. This may be required by external authorities during inspections and audits. External assets such as ERRV, close by installations, helicopters, coastguards, health services and shore based management may also be involved in drills to simulate a comprehensive involvement of all entities that may be required to provide support in a real time emergency. Each duty holder will have an established shore based Emergency Control Centre (ECC) which will be manned by personnel having access to the highest decision-making personnel so that ECC can be called upon at the discretion of OIM when assistance is required. In addition, ECC's function includes:

- maintaining communication with OIM and other shore based parties to ensure availability of resources to evacuate personnel or coordinate other assistance as required;
- ensuring list of POB is available so that it could be utilised fully to inform all relevant parties;

- keeping all parties informed about the situation. These include:
  - various government agencies such as Maritime and Coastguard Agency, Civil Aviation Authority, Police, HSE;
  - senior management;
  - press and media;
  - families and friends of workers on-board the installation.

As recommended by HSE,[1] the following is a list of exercises that must be carried out during drills:

- musters and use of alternative muster points;
- location and handling of casualties;
- fire fighting;
- MOB;
- helideck incident;
- helicopter incident;
- well incident;
- hydrocarbon production incident;
- ship collision incident;
- preparation for evacuation;
- preparation for abandonment.

7.4 Offshore Search and Rescue (SAR) and Emergency Organisation

Various organisations in the oil and gas industry have set up SAR systems to cope with maritime emergencies. This system is mainly set up under the International Convention on Maritime Search and Rescue (SAR) which was adopted on 27 April 1979 and entered into force on 22 June 1985. According to the SAR convention, the world is divided into 13 SAR areas within which relevant countries are responsible for SAR. Within this convention, governments of the countries are required to:

- establish a legal framework to implement the requirements;
- nominate a responsible authority;
- make available the required resources;
- establish communication facilities;
- establish coordination of all entities involved including provision of services for international cooperation.

The IAMSAR (International Aeronautical and Maritime Search and Rescue) manual resulting from the SAR convention describes the requirements and international standards to be followed for all aspects of SAR.

In the UK, EPOL (Emergency Preparedness Offshore Liaison Group)[2] was formed in 1995 as a result of the Piper Alpha disaster in

1988. In 2009, EPOL led to the development of OERWG (Offshore Emergency Response Working Group) that acts as a central body for bringing together emergency services, regulators, helicopter operators and duty holders/operators/owners of the offshore installations. This group reviews the SAR procedures for the local oil and gas industry in the UK Search and Rescue Region (UKSRR) and are published in the document 'Integrated Offshore Emergency Response'.

According to the PFEER regulations, the duty holder must ensure effective arrangements for the recovery of persons following an escape or evacuation from the installation, rescue of persons near the installation and then taking such persons to a place of safety. In order to fulfil this obligation, the entities described below are involved.

7.4.1 HM Coastguard (HMCG)

In the UK, the Department for Transport (DfT), formerly known as the Department of Transport (DoT), provides the necessary mechanism for SAR functions within UKSRR. HM Coastguard is a branch of the MCA which is a DfT agency. As per the SAR convention, the UK is required to have MRCCs (Maritime Rescue Coordination Centres) to coordinate any SAR operations. MRCCs may be assisted by MRSCs (Maritime Rescue and Coordination Sub-centres). In the UK, MRCCs are located in the Shetland Islands, Aberdeen, Humber and Liverpool.

The UK's MRCCs and MRSCs are supported by ARCC (Aeronautical Rescue Coordination Centre) located at RAF Kinloss which also provides one Nimrod MPA (Maritime Patrol Aircraft) for long range assistance. This aircraft is kept in a state of readiness at two hours notice. ARCC controls the deployment of helicopters from RAF, RN and HM coastguards. In addition, help can also be obtained from RNLI lifeboats which are located around the coast.

HMCG has an overall function of SAR coordinator within the UKSRR. HMCG is divided into nine SAR areas within which there are two or more Coastguard Rescue Teams (CRTs) that specialise in coastal search techniques and surveillance. HMCG stations maintain a continuous watch on VHF, MF and DSC distress channels, AIS (Automatic Identification System) and satellite communications. Distress satellite messages are received in MRCC Falmouth and routed to the required SAR facilities.

7.4.2 Police

The police in the UK are given criminal jurisdiction for oil and gas installations within the UKCS. However, unlike their role in shore based emergencies where they coordinate the response to any emergency, they are not required to coordinate the response to offshore emergencies. Instead, HM Coastguard will carry out this function for offshore civil emergencies. There are certain scenarios where police will be involved such as MOB, missing person, terrorist incident or threat, unexpected death or protesters boarding an installation. For other incidents, police will be notified for information only so that they could be called upon if any assistance is required.

7.4.3 RNLI

The Royal National Lifeboat Institution (RNLI) is a charitable organisation in the UK which was set up to assist in saving lives at sea and inland waters. RNLI maintains a fleet of hovercraft, all weather lifeboats and personnel trained to respond to any maritime emergencies.

7.4.4 Merchant Ships

Ships at sea are legally obliged to offer assistance to persons in distress. HM Coastguard/MCA will direct any ships in the vicinity to assist in the SAR operations. MCA charters four Emergency Towing Vessels (ETVs) placed strategically around the UK coast to provide assistance as required.

7.4.5 On Scene SAR Coordination

As per the IAMSAR requirements, each SAR operation is required to have the following nominated entities:

- SCs (SAR coordinators): nominated by the estate, SCs have an overall responsibility for coordination of SAR.
- SMC (SAR Mission Coordinator): an SMC is responsible for management of a SAR operation usually at an MRCC.
- OSC (On Scene Commander): when more than one SAR facility is operating in a SAR operation, one of them is nominated as an OSC. This may be a SAR unit, an aircraft, HMCG, the OIM of the installation requiring assistance or the master of a vessel assisting the SAR operations.
- CSS (Coordinator Surface Search): this role is assigned to the most suitable vessel that can carry out surface search for survivors.
- SRU (SAR Unit): an SRU is a facility supplied with suitable equipment and trained personnel to conduct SAR operations.
- Reception platform: when survivors are rescued from the casualty platform or sea they will be transferred to a reception platform which can be an ERRV or another neighbouring platform. Normally an installation's ERP would have

defined the reception platform to avoid confusion at a later time.

7.5 Abandoning an Offshore Installation

PFEER regulations require duty holders to identify and provide means for the safe evacuation of personnel to a place of safety. In order to meet this requirement, a number of methods are available to evacuate or escape from offshore installations. These methods are categorised into three areas below and are described in this section:

Figure 7.4 Scramble net (Source: http://billypugh.com)

1. Primary means of evacuation – helicopters.
2. Secondary means of evacuation – TEMPSC.
3. Tertiary means of escape – scramble net, knotted rope, personal descent devices.

7.5.1 Helicopters

Helicopters provide a vital role not only in transporting personnel to the offshore platforms but also when they need to be evacuated. The OIM may call upon helicopter support to evacuate non-essential personnel as soon as a warning for an emergency is received.

Once on the muster station, personnel will be given specific guidance for evacuation by using helicopters. Depending upon circumstances, normal helicopter boarding may take place, but if the installation has suffered damage due to fire/smoke, helicopter winching may be deployed to evacuate personnel from the designated areas.

7.5.2 Support Vessel

An offshore support vessel or ERRV (commonly referred to as a standby vessel) can be boarded by use of personnel transfer baskets with cranes. If circumstances don't permit the use of transfer baskets, escape chutes, ladders, scissor stairs or other tertiary means of escape could be used to reach the water level and then board the support vessel.

7.5.3 Tertiary Escape

Knotted escape ropes, ladders, chutes, scramble net, personal descent devices are all collectively referred to as tertiary escape devices.

Figure 7.5 Knotted escape rope (Source: http://billypugh.com)

7.5.3.1 Escape Chute

Escape chutes are used as a tertiary means of escape that allow rapid evacuation from the decks of high structures such as offshore installations. The total length of the chute also factors in horizontal movement due to pressure of wind.

The length of a chute depends upon the height of the deck where it is installed. It is then divided into cells by use of slanting dividers to control the speed of descent through the zig-zag direction of the dividers. Support is provided by steel rings at 1 m lengths throughout the chute. These rings are supported by steel wire ropes.

A chute's body is made from fire-retardant web with openings at various predetermined levels to allow escape or entry at different heights. Normally, the container within which a chute is kept is constructed of fire/blast proof materials. Design features incorporate sufficient dimensions of all components so that personnel can descend through the chute whilst wearing lifejackets, abandon suits or breathing apparatus. A descent rate of 10–30 persons per minute depending upon the drop can be achieved by escape chutes.

A chute can be deployed simply by operating a pneumatic cutter or a knife that releases the securing straps and thereby allowing the chute to drop down under gravity. The speed of descent can be controlled by use of brakes in some cases.

As the chute reaches the water level, the liferafts stowed at its base are inflated automatically. Some manufacturers provide a landing raft which is connected to additional liferafts on its side. The landing raft inflates automatically in all cases but the additional liferafts may need to be inflated by the first person descending through the chute.

Some variations in the design can be found between different manufacturers below:

- Viking life
- Survitec/Risk Safety Systems, USA.

7.5.4 Miscellaneous Means of Escape

Many installations are connected to accommodation/floatels via a gangway or link-bridge. Provided the floatel's safety case establishes it is safe to do so, the gangway can be dislodged after all personnel have transferred to the floatel. This may only be used to buy time to escape using other means described earlier.

7.5.5 Survival Craft

Two types of survival craft are supplied on the offshore installations. These are lifeboats and liferafts. Both types of survival craft can be used for dry evacuation to the surface of water from where survivors can be picked up by helicopter, ERRV or other vessels. Further detail of the types of survival craft, their launching mechanism, etc. is given in the next section.

Ships and offshore installations are required by law to carry specific life-saving appliances which include lifeboats or liferafts. Their purpose is to provide a means of escape from the parent vessel when it has to be abandoned.

Lifeboats vary in size, shape and carrying capacity. Generally, their size depends upon the number of personnel employed on an installation. Even on some of the largest installations or ships, the maximum POB may be restricted due to the limit on the survival craft capacity.

The sinking of the *Titanic* in 1912 caused a loss of life for over 1500 personnel, mainly because the 'lifeboat' capacity was less than the number of personnel carried. The *Titanic* incident appears to be the most significant event in the development of SOLAS – the International Convention on Safety of Life at Sea. The lifeboats are required to comply with IMO's LSA code. As the technology developed, larger workforces continued to be employed on ships and offshore installations and hence the capacity of lifeboats continued to increase. Since the SOLAS convention provided a good base for standards relating to design of all survival craft, the offshore industry complies with these standards.

On modern installations of today, lifeboats are referred to as Totally Enclosed Motor Propelled Survival Craft (TEMPSC). As the name suggests, these are totally enclosed to provide protection from the elements to the occupants and fitted with an inboard engine to provide a means of propulsion. Traditionally launched and recovered by davits, a further development in the launching mechanism has been the development of free-fall launching arrangements. Most of today's lifeboats are built from glass fibre reinforced polyester material (GRP). The next sections discuss these areas to give the reader knowledge of lifeboats.

Offshore installations are required to provide a minimum lifeboat capacity of 150 per cent of the POB. Additional lifeboats are required for areas where personnel may not be able to reach the lifeboat in a short time period such as from a temporary refuge. For any tasks involving dive work, hyperbaric lifeboats (described later in the chapter) must also be provided on the basis of safety case requirements.

There are two general types of lifeboats found on offshore installations. These are

1. TEMPSC – Totally Enclosed Lifeboats launched by single or twin falls;
2. Free-fall – also totally enclosed but different in design due to their free-fall launch mechanism.

Figure 7.6 A lifeboat being tested in water during a drill

Figure 7.7 Inside view of a TEMPSC

7.5.5.1 Lifeboat Construction Characteristics

Whilst lifeboats in the offshore industry are not strictly required to comply with IMO/SOLAS requirements, there being no international alternative generally all lifeboats will be approved to at least SOLAS standards. The main structural features of all types of lifeboat are reproduced here for the reader's convenience.

- Lifeboats are made of a rigid fire-retardant or non-combustible hull.
- They are capable of maintaining positive stability when in an upright position in calm water and loaded with their full complement of persons and equipment and holed in one location below the waterline assuming no loss of buoyancy.
- They are capable of being safely launched under all conditions of trim of up to 10° and list of up to 20° either way.
- They are capable of being launched and towed when the ship is making headway at a speed of 5 knots in calm water. This requirement applies to the 'non-fixed' structures such as MODU and ships but the lifeboats are built to the same standards, hence all lifeboats comply with this requirement.
- Seating is provided on thwarts, benches or fixed chairs that can support weight of up to 100 kg.
- Except for free-fall lifeboats, each lifeboat to be launched by falls will have sufficient strength to withstand:
  - a load of 1.25 times the total mass of lifeboat with its full complement for metal hull boats and 2 times the total mass for boats of non-metal hulls.

- a lateral impact against the ship/installation side at a velocity of at least 3.5 m/s and drop into water from a height of at least 30 m.
- Each seating position is clearly indicated in the lifeboat.
- Lifeboat access is arranged in such a way that it can be boarded within 3 minutes from the time the instruction to board is given.
- A boarding ladder providing accesses to 0.4 m below the lifeboat's light waterline shall be provided.
- The lifeboat is designed to ensure helpless people can be brought on-board either from the sea or on stretchers.

7.5.5.2 Lifeboat Propulsion

As the name suggests, TEMPSC (Totally Enclosed Motor Propelled Survival Craft), are fitted with an inboard engine with the minimum specifications stated below:

- provided with a compression ignition engine capable of operating with a fuel with flashpoint of more than 43°C;
- the engine provided with either a manual starting system, or a power starting system with two independent rechargeable energy sources. Any necessary starting aids shall also be provided. The engine starting systems and starting aids shall start the engine at an ambient temperature of –15°C within 2 minutes of commencing the start procedure;
- the engine capable of operating for not less than 5 minutes after starting from cold with the lifeboat out of the water;
- the engine capable of operating when the lifeboat is flooded up to the centre line of the crankshaft;
- the propeller shafting so arranged that the propeller can be disengaged from the engine. Provision shall be made for ahead and astern propulsion of the lifeboat;
- the exhaust pipe so arranged as to prevent water from entering the engine in normal operation;
- all lifeboats designed with due regard to the safety of persons in the water and to the possibility of damage to the propulsion system by floating debris;
- the speed of a lifeboat when proceeding ahead in calm water, when loaded with its full complement of persons and equipment and with all engine powered auxiliary equipment in operation, at least 6 knots and at least 2 knots when towing the largest liferaft carried and loaded with its full complement of persons and equipment or its equivalent;
- sufficient fuel, suitable for use throughout the temperature range expected in the area in which the ship operates, provided to run the fully loaded lifeboat at 6 knots for a period of not less than 24 h;

- the lifeboat engine, transmission and engine accessories enclosed in a fire-retardant casing or other suitable arrangements providing similar protection;
- adequate means provided to reduce the engine noise so that a shouted order can be heard;
- starter batteries provided with casings which form a watertight enclosure around the bottom and sides of the batteries; the battery casings have a tight fitting top which provides for necessary gas venting;
- the lifeboat engine and accessories designed to limit electromagnetic emissions so that engine operation does not interfere with the operation of radio life-saving appliances used in the lifeboat;
- means provided for recharging all engine starting, radio and searchlight batteries; radio batteries shall not be used to provide power for engine starting;
- water resistant instructions for starting and operating the engine provided and mounted in a conspicuous place near the engine starting controls.

7.5.5.3 Lifeboat Fittings

In order to comply with SOLAS requirements, lifeboats must be fitted with the following accessories:

- at least one drain valve (with a cap attached to the lifeboat by a lanyard, a chain) fitted near the lowest point in the hull, which shall automatically open to drain water;
- all lifeboats provided with a rudder and tiller. When a wheel or other remote steering mechanism is also provided the tiller shall be capable of controlling the rudder in case of failure of the steering mechanism;
- except in the vicinity of rudder and propeller, suitable handholds provided or a buoyant lifeline shall be becketed around the outside of the lifeboat above the waterline and within reach of a person in the water;
- lifeboats which are not self-righting when capsized shall have suitable handholds on the underside of the hull to enable persons to cling to the lifeboat;
- all lifeboats shall be fitted with sufficient watertight lockers or compartments to provide for the storage of the small items of equipment, water and provisions;
- every lifeboat to be launched by a fall or falls, except a free-fall lifeboat, shall be fitted with an on-load/off-load release mechanism;
- every lifeboat shall be fitted with a device to secure a painter near its bow;
- lifeboats intended for launching down the side of a ship shall have skates and fenders;

- a manually controlled exterior light shall be fitted. The light shall be white and be capable of operating continuously for at least 12 h with a luminous intensity of not less than 4.3 candela (cd) in all directions of the upper hemisphere. However, if the light is a flashing light it shall flash at a rate of not less than 50 flashes and not more than 70 flashes per min for the 12 h operating period with an equivalent effective luminous intensity;
- a manually controlled interior light shall be fitted inside the lifeboat capable of continuous operation for a period of at least 12 h;
- every lifeboat shall be so arranged that an adequate view forward, aft and to both sides is provided from the control and steering position for safe launching and manoeuvring;
- a lifeboat with a self-contained air support system shall be so arranged that, when proceeding with all entrances and openings closed, the air in the lifeboat remains safe and breathable and the engine runs normally for a period of not less than 10 minutes. During this period the atmospheric pressure inside the lifeboat shall never fall below the outside atmospheric pressure nor shall it exceed it by more than 20 hPa. The system shall have visual indicators to indicate the pressure of the air supply at all times;
- a fire-protected lifeboat when waterborne shall be capable of protecting the number of persons it is permitted to accommodate when subjected to a continuous oil fire that envelops the lifeboat for a period of not less than 8 minutes.

7.5.5.4 Lifeboat Equipment

All items of lifeboat equipment shall be secured within the lifeboat by lashings, storage in lockers or compartments, storage in brackets or similar mounting arrangements or other suitable means.

Figure 7.8 Lifeboats on a jack-up rig

However, in the case of a lifeboat to be launched by falls the boat-hooks shall be kept free for fending off purposes. The equipment shall be secured in such a manner as not to interfere with any abandonment procedures. All items of lifeboat equipment shall be as small and of as little mass as possible and shall be packed in a suitable and compact form.

The normal equipment of every lifeboat shall consist of:

1. except for free-fall lifeboats, sufficient buoyant oars to make headway in calm seas. Thole pins, crutches or equivalent arrangements shall be provided for each oar provided. Thole pins or crutches shall be attached to the boat by lanyards or chains;
2. two boat-hooks;
3. a buoyant bailer and two buckets;
4. survival manual;
5. an operational compass which is luminous or provided with suitable means of illumination. In a totally enclosed lifeboat, the compass shall be permanently fitted at the steering position; in any other lifeboat, it shall be provided with a binnacle and suitable mounting arrangements if necessary to protect it from the weather;
6. A sea anchor of adequate size fitted with a shock resistant hawser which provides a firm hand grip when wet. The strength of the sea anchor, hawser and tripping line shall be adequate for all sea conditions;
7. Two efficient painters of a length equal to not less than twice the distance from the stowage position of the lifeboat to the waterline in the lightest seagoing condition or 15 m, whichever is the greater. On lifeboats to be launched by free-fall launching, both painters shall be stowed near the bow ready for use.

Figure 7.9 Lifeboats and rescue boat on a an offshore installation

On other lifeboats, one painter attached to the release device shall be placed at the forward end of the lifeboat and the other shall be firmly secured at or near the bow of the lifeboat ready for use;

8. two hatchets, one at each end of the lifeboat;

9. watertight receptacles containing a total of 3 litres of fresh water for each person the lifeboat is permitted to accommodate, of which either 1 litre per person may be replaced by a desalting apparatus capable of producing an equal amount of fresh water in 2 days, or 2 litre per person may be replaced by a manually powered reverse osmosis desalinator capable of producing an equal amount of fresh water in 2 days;

10. a rustproof dipper with lanyard;

11. a rustproof graduated drinking vessel;

12. a food ration totalling not less than 10,000 kilojoules for each person the lifeboat is permitted to accommodate; these rations shall be kept in airtight packaging and be stowed in a watertight container;

13. four rocket parachute flares;

14. six hand flares complying with the requirements;

15. two buoyant smoke signals complying with the requirements;

16. one waterproof electric torch suitable for Morse signalling together with one spare set of batteries and one spare bulb in a waterproof container;

17. one daylight signalling mirror with instructions for its use for signalling to ships and aircraft;

18. one copy of the life-saving signals prescribed by regulation V/16 on a waterproof card or in a waterproof container;

19. one whistle or equivalent sound signal;

20. a first aid outfit in a waterproof case capable of being closed tightly after use;

21. anti-seasickness medicine sufficient for at least 48 h and one seasickness bag for each person;

Figure 7.10 A TEMPSC in davits on a an offshore installation

22. a jack-knife to be kept attached to the boat by a lanyard;
23. three tin openers;
24. two buoyant rescue quoits, attached to not less than 30 m of buoyant line;
25. if the lifeboat is not automatically self-bailing, a manual pump suitable for effective bailing;
26. one set of fishing tackle;
27. sufficient tools for minor adjustments to the engine and its accessories;
28. portable fire-extinguishing equipment of an approved type suitable for extinguishing oil fires;
29. a searchlight with a horizontal and vertical sector of at least 6° and a measured luminous intensity of 2500 cd which can work continuously for not less than 3 h;
30. an efficient radar reflector, unless a survival craft radar transponder is stowed in the lifeboat;
31. thermal protective aids (TPA) sufficient for 10 per cent of the number of persons the lifeboat is permitted to accommodate or two, whichever is the greater.

7.5.5.5 Lifeboat Markings

Each lifeboat shall be fitted with a permanently affixed approval plate, endorsed by the flag state or administration or its representative, containing at least the following items:

- manufacturer's name and address;
- lifeboat model and serial number;
- month and year of manufacture;
- number of persons the lifeboat is approved to carry.

The name and port of registry of the ship/installation to which the lifeboat belongs shall be marked on each side of the lifeboat's bow in block capitals of the Roman alphabet.

Means of identifying the installation/ship to which the lifeboat belongs and the number of the lifeboat shall be marked in such a way that they are visible from above.

7.5.5.6 Additional Requirements for TEMPSC

TEMPSC are required to comply with the general requirements for lifeboats and, in addition, comply with the requirements given here. Every totally enclosed lifeboat shall be provided with a rigid watertight enclosure which completely encloses the lifeboat. The enclosure shall be so arranged that:

1. it provides shelter for the occupants;
2. access to the lifeboat is provided by hatches which can be closed to make the lifeboat watertight;

Figure 7.11 A TEMPSC being launched into water

3. except for free-fall lifeboats, hatches are positioned so as to allow launching and recovery operations to be performed without any occupant having to leave the enclosure;
4. access hatches are capable of being opened and closed from both inside and outside and are equipped with means to hold them securely in open positions;
5. except for a free-fall lifeboat, it is possible to row the lifeboat;
6. when the lifeboat is in the capsized position with the hatches closed and without significant leakage, it is capable of supporting the entire mass of the lifeboat, including all equipment, machinery and its full complement of persons;
7. it includes windows or translucent panels which admit sufficient daylight to the inside of the lifeboat with the hatches closed to make artificial light unnecessary;
8. its exterior is of a highly visible colour and its interior of a light colour which does not cause discomfort to the occupants;
9. handrails provide a secure handhold for persons moving about the exterior of the lifeboat, and aid embarkation and disembarkation;
10. persons have access to their seats from an entrance without having to climb over thwarts or other obstructions;
11. during operation of the engine with the enclosure closed, the atmospheric pressure inside the lifeboat shall never be above or below the outside atmospheric pressure by more than 20 hPa.
12. Capsizing and re-righting:
    a. Except in free-fall lifeboats, a safety belt shall be fitted at each indicated seating position. The safety belt shall be designed to hold a person with a mass of 100 kg securely in place when the lifeboat is in a capsized position. Each set of safety belts for a seat shall be of a colour which contrasts with the belts for seats immediately adjacent. Free-fall

lifeboats shall be fitted with a safety harness at each seat in contrasting colour designed to hold a person with a mass of 100 kg securely in place during a free-fall launch as well as with the lifeboat in capsized position.

b. The stability of the lifeboat shall be such that it is inherently or automatically self-righting when loaded with its full or a partial complement of persons and equipment and all entrances and openings are closed watertight and the persons are secured with safety belts.

c. The lifeboat shall be capable of supporting its full complement of persons and equipment when the lifeboat is in the damaged condition and its stability shall be such that in the event of capsizing, it will automatically attain a position that will provide an above-water escape for its occupants. When the lifeboat is in the stable flooded condition, the water level inside the lifeboat, measured along the seat back, shall not be more than 500mm above the seat pan at any occupant seating position. The design of all engine exhaust pipes, air ducts and other openings shall be such that water is excluded from the engine when the lifeboat capsizes and re-rights.

13. Propulsion:

a. The engine and transmission shall be controlled from the helmsman's position.

b. The engine and its installation shall be capable of running in any position during capsize and continue to run after the lifeboat returns to the upright or shall automatically stop on capsizing and be easily restarted after the lifeboat returns to the upright. The design of the fuel and lubricating systems shall prevent the loss of fuel and the loss of more than 250ml of lubricating oil from the engine during capsize.

c. Air-cooled engines shall have a duct system to take in cooling air from, and exhaust it to, the outside of the lifeboat. Manually operated dampers shall be provided to enable cooling air to be taken in from, and exhausted to, the interior of the lifeboat.

7.5.6 Launching a Lifeboat

For each davit, a plate showing the instructions to launch the lifeboat is installed near the embarkation deck. These instructions are specifically written for the equipment installed on any ship and show the exact sequence for launching the lifeboat. With a qualified lifeboat coxswain in charge, the general procedure given below should be followed:

1. Remove boat lashings.
2. Remove all protection covers.

3. Second person to release slip hook located in the middle of the davit.

4. Ensure:
   a. harbour safety pins (if fitted for boats on offshore structures) are released and stowed in safe location;
   b. launch area is clear;
   c. electrical supply is disconnected.

5. One person to enter the boat and plug the drains.

6. Check if the maintenance pendant has been removed.

7. Check and clear any obstructions.

8. Check tricing-in pendants and bowsing-in tackle (not fitted on boats that can be embarked from stowage position and launched straight into water).

9. Physically check the hooks and on-load release gear to ensure everything is in order.

10. Fall Prevention Device (FPD) should be fitted to both bow and stern hooks and falls. FPDs may not be required for new design release gear.

11. Lower the boat to sea level to ensure all components operate normally. In training and drills, hoist the boat back to embarkation position to ensure all components operate satisfactorily. In emergencies, this will not be required.

12. Once the boat is ready to be launched, upon orders from the OIM:
    a. Ensure all personnel are wearing lifejackets and suitable PPE.
    b. Other designated personnel can board the boat.
    c. Check engine is ready to start.
    d. Check fuel valve is set for operational start-up.
    e. Request permission to start engine from person in charge. Switch on for 5 seconds, and then stop engine. This step may vary for different manufacturers.
    f. All hatches should now be closed and secured, then all personnel in the boat should strap into seating using the seatbelts provided. This is then confirmed to the person in charge of launching the boat by VHF.
    g. Once all persons have been strapped, the OIM will confirm all clear to launch.

13. Ensure:
    a. over side floodlights (if fitted) are moved out of the lifeboat/davit's launch path to prevent damage;
    b. any boat lashings are removed;
    c. battery charging cable is removed.

14. All actions must be taken after obtaining approval from the OIM/master.

15. If launching from within the craft:
    a. Pull firmly on the remote control grip of the brake release wire to start lowering the boat. Once properly activated the lowering will continue until the boat is waterborne.

b. Boat is then lowered by using the brake release wire from inside the lifeboat. The brake release wire should never be wrapped around hands or arms.

c. Once the boat is about 1 m above the waterline the person in charge on the deck of the installation will communicate this to the lifeboat coxswain who will pause the launch to ensure that the strop type FPDs are removed whilst there is still tension on the fall wires, preventing injury from loose fall wires. Boat is then lowered to the water by using the brake release wire from inside the lifeboat. Tension should be kept on the brake wire until after the boat is fully waterborne to ensure the falls become slack.

d. During drills, engine should be started when the boat is fully waterborne but in emergency evacuation, engine should be started before embarking passengers.

e. The fall hooks can now be released from inside the lifeboat ensuring the hook release interlock is in 'off' or cleared position after which, the lever can be operated to release both hooks simultaneously. If the release interlock does not disengage automatically, the emergency override must be used.

f. Once the engine is started, gear engaged, the painter line can be cast off. Once all lines are gone this is confirmed to the coxswain who then manoeuvres the boat away from the ship's side/installation.

7.5.6.1 Launching a Lifeboat in Rough Weather

Many life threatening accidents have occurred during drills when lifeboats were being launched or recovered. During launch, the lifeboat may impact the sides of the distressed vessel or installation and become severely damaged, and the occupants may fall into the sea suffering injury and even death. The complications of launching a lifeboat in rough weather may be the same for a ship or an installation; hence the reference is only made to ships.

Launching the lifeboats may be impossible if the parent vessel is listing significantly or if the falls become tangled. Once the lifeboat is waterborne, it may be unable to move away from the distressed vessel because high seas and wind continually push the lifeboat to the vessel or because of a malfunction in the propulsion system. This situation is even more dangerous during a fire or when the potential for an explosion exists.

Whilst the basic procedure for launching a conventional lifeboat remains the same, in rough seas, attention needs to be paid to the movement of waves and direction of wind. If practical, the lifeboat should be lowered into the trough of a wave. As the crest of the wave pushes the boat up, the on-load release mechanism should be operated to release both the falls. The person on the winch should

pay particular attention to the slack in the fall wires which may get entangled when very slack and, on the other hand, may cause other problems when the boat suddenly loses upthrust from water.

This will enable the boat to fall down into the next trough clearing it from the floating block and the falls. Furthermore, the drop into the trough will move the boat away from the ship. This move can be supplemented by using the full force of the engines to steer the boat away from the ship. Many injuries have occurred during this time when personnel have been struck by the floating block. Therefore it is paramount that personnel are not allowed to move out of the boat at any time until the boat is clear of the ship.

7.5.7 Starting Lifeboat Engine

Lifeboats are generally fitted with inboard diesel engines. Whilst the make and model of the engine may make some variations to the starting procedure, generally the following steps will apply:

1. Switch on the main switch.
2. Put the gear in neutral position.
3. Put the switch into 'run' position.
4. Push start button until the engine starts. The starter motor should not be required to work for more than 10–15 seconds continuously.

The engine can be stopped by pushing the STOP button. Once stopped, put the switch to the OFF position. For a cold start (below 15 degrees Celsius), special instructions need to be followed for the type of engine as some engines may be fitted with internal heaters to warm the engine. If this is the case, then the manufacturer's relevant instructions must be complied with concerning the heating time required for the engine.

7.5.7.1 Manual Starting Procedure

If the engine does not start through the electric starter motor due to the battery being flat or for any other reasons, then each engine is provided with a manual start system that can be used as below:

1. Put gear into neutral position.
2. Put handle into crank claw.
3. Lift decompression lever.
4. Turn the handle clockwise as quickly as possible, release the decompression lever but keep turning the handle until the engine starts.
5. In cold weather it is easier to start the engine after cranking it with the activated decompression lever before following the electric starting procedure.

7.5.8 Lifeboat Hook Release Gear

Different types of hook release gear are fitted on modern lifeboats. The objective of this gear is to release the lifeboat from the falls allowing it to manoeuvre away from the parent vessel or installations.

The differences in their designs and other characteristics are discussed below.

7.5.8.1 Off-load Release Gear

As the name suggests, the off-load release gear releases the lifeboat from the falls after the load of the boat is transferred to the water when it is waterborne. Disengaging the falls from the hook can also be carried out manually if the hooks are not automatically released after the boat weight transfers to the water.

This mechanism is known to cause difficulties during release when the boat is being lowered in rough seas or the vessel is making headway. In such cases, the weight may come off one fall whereas the other may still be under load, making it difficult for the falls to be released simultaneously. In many cases, the boat is required to be manoeuvred to remove the load from the falls and detach the boat. This may not be achieved due to weather and/or other factors stated here. As the time is critical to escape especially when the parent ship is sinking, the on-load release gear was invented to overcome off-load gear problems.

The loss of the semisubmersible floatel 'Alexander L. Kielland' in the Ekofisk field of the Norwegian sector in March 1980 drew attention to the difficulty in releasing TEMPSC from their falls and manoeuvring them away from the installation in adverse weather conditions. Stemming from this incident the need for lifeboat/ TEMPSC 'on-load' release gear was highlighted throughout the marine and offshore industries. Rather than relying on the craft to be waterborne with no weight on the falls to permit the hooks to be released, the new mechanism i.e. on-load release gear enables a boat to be released while its full weight is still supported by the falls, even if it is above the water. On-load release gear is now commonly known as 'Lifeboat Release and Retrieval System (LRRS)' or 'On-load LRRS (OLRRS)'.

7.5.8.2 On-Load Release Gear

Most of the TELBs (Totally Enclosed Lifeboats) except free-fall boats and PELBs (Partially Enclosed Lifeboats) are now fitted with on-load hook release gear. This system is designed to release the lifeboat from fall wires even when they are under the full load of the boat and its complement.

Even though the on-load release gear allows the boat to be released when the boat is at any height above the water surface, this *should never be attempted to avoid injury* to boat occupants.

IMO adopted new rules in order to enhance lifeboat safety, and new requirements for on-load release hooks entered into force on 1 January 2013. For new build vessels, the requirements apply to ships whose keel was laid after 1 January 2013. Vessels in operation shall comply with the requirements given in the amended SOLAS Reg. III/1.5 (Res. MSC.317 (89)) by their first dry docking after 1 July 2014, and at the latest by 1 July 2019.

7.5.8.3 On-Load Release Gear Features

1. **Hydrostatic Safety Interlock.** This interlock is made from a flexible membrane diaphragm enclosed in a housing above the lifeboat keel. When the boat is submerged, water enters through the inlets filling the cavity around the housing and thereby forcing the membrane upwards. A piston located above the membrane surface is forced upwards as a result and releases the mechanical interlock which otherwise prevents the operation of the release handle. A connecting cable allows the motion to be transferred to the indicator. As the membrane rises, the indicator moves from the locked to the unlocked sector of the release unit. This change can be seen by the coxswain in the lifeboat who can then pull the handle to operate the release gear.
2. **Recovering a Lifeboat.** During drills, the on-load release gear along with the hooks will be reset prior to the boat coming under the fall blocks. When this is done, the indicator remains in the same position. After attaching the fall wires to the boat, when the lifeboat is being recovered, hoisting should stop when the

boat is above the water surface so that the cavity has emptied and the interlock indicator has reset itself into the locked position. The coxswain should ensure that this procedure is carried out carefully as otherwise there is a danger of the release handle being operated during the recovery of the lifeboat leading to fatal consequences.

3. **Release Unit.** The release unit comprises the release handle, safety pin or lock, and indicator to denote operational status. The indicator is in the locked position i.e. the lifeboat is clear of the water. Under normal conditions, it should not be possible to operate the release handle for reasons explained earlier.

4. **Hook Assembly.** The hook assembly consists of the components fitted at the fore and aft ends of the boat to support its full weight when suspended from fall wires. The hook securing mechanism can be in the form of a cam or cone and is fitted inside the hook assembly. This must be reset once the release mechanism has been operated in order to allow the hook to be re-secured during recovery of the lifeboat. The hook assemblies have side plates which are in turn bolted to the main keel foundation.

5. **Cables.** The cables connect various components of the system. For example, the action of operating the hook release handle is transmitted through these cables which in turn cause the hook securing mechanism to rotate and release the hooks. Normal inspections should include these cables for any signs of damage because once these cables are damaged, they can compromise the whole mechanism.

6. **Emergency Interlock Override.** If for any reason, the release handle is found to be non-operational when the boat is waterborne, then emergency interlock override can be used to release the lifeboat from the falls. This can be achieved by breaking the protective cover on the indicator and manually moving it to the unlocked sector and then operating the release handle. However, this should never be attempted when the boat is NOT waterborne.

7.5.9 Fall Prevention Devices (FPDs)

FPDs[3] are secondary devices fitted between the lifeboat and the falls and are required for on-load release gear to provide additional safety during the launching and/or hoisting of a survival craft. This requirement applies to the lifeboat on-load release systems until they are compliant with the revised requirements given in the LSA code. Although on-load release systems fitted to lifeboats are safe if operated and maintained properly, there have been a number of accidents during drills and servicing. Many of these accidents were attributed to lack of maintenance, poor design or inadequate training. Failures of equipment can result in the premature opening of the on-load hook mechanism, causing the lifeboat to fall from the davits unexpectedly even with safety interlocks provided for in the design arrangements.

FPDs can be either pins or strops fitted to the on-load release hooks to prevent the lifeboat from falling to the water in the event of an equipment failure or accidental release. Fitting FPD should not involve any modification to the existing structure such as for pins. However, fitting of strops, etc. is allowed provided the SWL of each component is checked against the SWL required for the approved components. The system, including FPD, should be verified at the initial survey for a new ship or an equivalent survey for an existing ship if equipment is changed or modified. This is to ensure that the installed system with the FPD fitted functions correctly.

The ship's Master or the officer in charge of any lifeboat lowering or lifting operation should ensure that the lifeboat FPD are fitted before commencing any drill, testing, inspection or maintenance where persons are in the lifeboat, unless the lifeboat has either an off-load hook system or has been approved to be used without an FPD. The ship's crew must be fully trained in the operation of the FPD fitted to the lifeboat on their ship. The procedure to be followed should be contained in ISM documentation and the ship's training manual.

7.6 Free-fall Lifeboats

When abandoning a ship or an offshore installation, escape needs to be fast and safe. Generally, the traditional lifeboat/davit combination takes more time. The speed at which the lifeboat can be boarded and launched is extremely significant in determining the chances of safe escape.

A solution for this problem was identified in the form of free-fall lifeboats. These lifeboats are completely different from traditional lifeboats in that their launching arrangement is different, they are installed with their forward end pointing to the sea and in contrast to 100 per cent capacity on each side of the ship for other TEMPSC, only one lifeboat accommodating 100 per cent of complement is required.

The hoisting and gravity lowering is operated through a specially designed davit system which uses two double wires ropes attached to the lifting eyes on the boat.

Free-fall lifeboats need to comply with the requirements of TELB and in addition the requirements given in this section.

Each free-fall lifeboat shall be of sufficient strength to withstand, when loaded with its full complement of persons and equipment, a free-fall launch from a height of at least 1.3 times the free-fall certification height. This type of lifeboat shall make positive headway immediately after water entry and shall not come into contact with the ship/installation after a free-fall launching against a trim of up

Figure 7.13 Two free-fall lifeboats on an offshore installation

to 10° and a list of up to 20° either way from the certification height when fully equipped and loaded with:

1. its full complement of persons;
2. occupants so as to cause the centre of gravity to be in the most forward position;
3. occupants so as to cause the centre of gravity to be in the most aft position; and
4. its operating crew only.

For oil tankers, chemical tankers and gas carriers with a final angle of heel greater than 20° calculated in accordance with the International Convention for the Prevention of Pollution from Ships, 1973, a lifeboat shall be capable of being free-fall launched at the final angle of heel and on the basis of the final waterline of that calculation.

7.6.1 Free-fall Lifeboat Fittings

Generally the free-fall hook is located at the stern of the boat. The primary release device is in the boat cabin next to the helmsman and is controlled by them. An additional secondary release device, usually located aft in the cabin, is also provided. These two systems are kept independent of each other to provide back-up in case one system fails.

Each free-fall lifeboat shall be fitted with a release system which shall:

1. have two independent activation systems for the release mechanisms which may only be operated from inside the lifeboat and be marked in a colour that contrasts with its surroundings;
2. be so arranged as to release the boat under any condition of loading from no load up to at least 200 per cent of the normal load caused by the fully equipped lifeboat when loaded with the number of persons for which it is to be approved;
3. be adequately protected against accidental or premature use;
4. be designed to test the release system without launching the lifeboat;
5. be designed with a factor of safety of 6 based on the ultimate strength of the materials used.

7.7 Big Persons in Lifeboats

Lifeboats approved to the SOLAS standard considered the average weight of a person to be 75kg. However, since 2005, it has been established through research that the average weight and size of

Figure 7.14 Inside view of a free-fall lifeboat

Figure 7.15 Free-fall lifeboat boarding arrangement

persons has increased resulting in a mismatch between the stated lifeboat capacity and the number of persons it can actually accommodate.

PFEER regulations in the UK require the duty holders to ensure the average weight and size of the persons occupying any lifeboat to be considered when determining its capacity. As a consequence, the duty holders may be required to either replace the lifeboats and associated gear with bigger ones, reduce the number of occupants in the lifeboats and therefore reduce the POB on a lifeboat or alternatively remove some equipment such as water rations, fuel or other equipment with the approval of the certifying authority. The complications of the size of persons need to be assessed by selecting a group of largest persons on-board and seating them in the lifeboat with their seat belts fastened. Some modifications to the components that do not fit will be required to ensure compliance with legislation as well as to ensure availability of the required craft for emergencies.

7.8 Single Fall TEMPSC

Now rarely used, single fall TEMPSC were found commonly in older offshore installations. They were also referred to as survival capsules mainly manufactured by Whittaker Corporation.

The design consisted of a spheroid with an access hatchway in the upper part and a heavy keel. Whilst the construction of a survival capsule is similar to a lifeboat, its shape and launching mechanism is different in that it is oval/circular in shape and utilises a single fall to launch/recover the capsule. The capsule is launched under gravity and hoisted by using an electric winch. This capsule is capable of escaping from an installation and does not have any directional steering capabilities due to its shape. Consequently, it cannot be used for mustering other craft such as liferafts.

Figure 7.16 Whittaker capsule

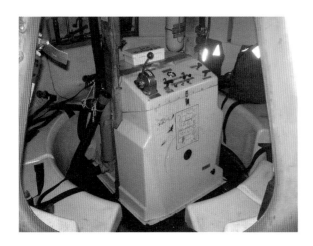

7.9 Hyperbaric Lifeboat

The suitability of various options for a safe hyperbaric evacuation depend on a number of factors including the geographical area of operation, environmental conditions and any available offshore or onshore medical and support facilities. Options available to the diving workers will include:

1. hyperbaric self-propelled lifeboats;
2. towable hyperbaric evacuation units;
3. hyperbaric evacuation units which may not be suitable for off-loading onto an attendant vessel;
4. transfer of the diving bell to another facility;
5. transfer of the divers from one diving bell to another when in the water and under pressure; and
6. negatively buoyant unit with inherent reserves of buoyancy, stability and life support capable of returning to the surface to await independent recovery.

Hyperbaric Lifeboats (HLB) are also called Self-Propelled Hyperbaric Lifeboats (SPHL). These boats are fitted with engines, generator and a life support system to provide survival capability of up to 72 hours for personnel rescued from a diving system. The life support system consists of gas supplies, breathing gas system, decompression equipment, environmental control system, heating or cooling and other equipment required to provide a safe environment for divers in the hyperbaric evacuation unit under all ranges of pressure that they may be exposed to during evacuation and during the decompression stages. Additional towing mechanisms may be fitted for this boat to be towed by a support or recovery vessel. MGN 83 'Guidelines and Specifications for Hyperbaric Evacuation Systems' provides full details of the hyperbaric systems.

7.10 Inflatable Liferafts

Liferafts are considered a back-up for TEMPSC. All liferafts are constructed in such a way that they can withstand exposure to the harshness of the sea for at least 30 days. In addition, for liferafts stowed at a height of 18 m above the water line, their design requires them to be capable of withstanding a drop from a height of 18 m. When abandoning from an offshore installation, the survivors are advised to board the survival craft dry. This can be achieved if they jump into the liferaft; hence the liferaft needs to be built to withstand these jumps. Other powered survival crafts are required to congregate liferafts by using the grablines fitted around them which need to be constructed to withstand a pull of up to 3 knots in calm water.

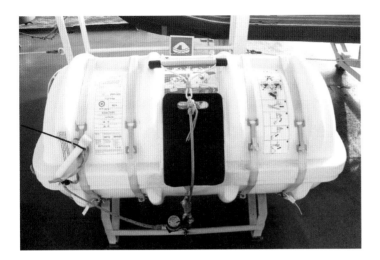

Figure 7.17 A davit launched liferaft
in cradle

There are three main types of liferafts:

1. throw over type
2. davit launched
3. aviation liferafts.

7.10.1 Launching a Deck Stowed Throw Over Type Liferaft

Liferafts can be fitted with a hydrostatic release unit to allow automatic inflation if the installation sinks without the liferaft being released first. When launching them manually, it is preferable to launch on the lee side to avoid drifting on to the ship or installation that may be on fire or sinking. This operation can require one, two or four persons depending upon the stowage position, size of the liferaft and the design of the cradle. Once this is established, the following procedure should be followed after receiving an order from OIM to launch it:

Figure 7.18 An inflated marine liferaft

Figure 7.19 Multiple liferafts combined stowage for simultaneous launch

1. Ensure the painter line is secured to a strong point on the deck.
2. Remove the lashing that secures the liferaft in its cradle or ramp (for stowage of multiple rafts in the same cradle).
3. Check the area below is clear of obstructions and other persons in waters (if any).
4. Lift the liferaft container to throw it overboard.
5. Pull the painter to its full length to trigger the CO_2 gas cylinder. Triggering may require a strong pull/jerk on the painter line to activate the gas release.
6. Once CO_2 is released, it will inflate the liferaft.

Once inflated, the liferaft can be boarded by the use of a boarding ladder or other means of access. First preference must be given to boarding it dry. If this is not possible, then jump from shortest possible height into the entrance of the liferaft otherwise jump into water following the procedure below:

- Ensure the area to jump in is clear from debris or other persons.
- Hold nose with one hand to avoid water entry.
- Hold lifejacket with the other hand.
- Keep feet together.
- Look straight ahead to avoid falling flat on water.
- Step forward and jump into water.

7.10.2 Raft Capsize

A liferaft may capsize due to various reasons including inverted inflation, capsize due to wind or waves. Regardless of how it inverts, it can be righted by one person by following the procedure below:

1. Climb on to liferaft cylinder whilst holding the righting strap.
2. With feet placed firmly on the cylinder, tilt the raft into wind.

Figure 7.20 Righting a capsized liferaft (Source: Warsash Maritime Academy)

3. Pull back on the righting strap by putting body weight on the righting strap.
4. The raft will follow you whilst you drop in water on your back.
5. To avoid being trapped under the liferaft, it is preferred to fall on back and use righting strap as a guideline to pull out.

Empty liferafts are quite prone to capsizing. However, fully manned liferafts may also capsize but this can be avoided by streaming the drogue at an early stage to provide it some stability in addition to the stability provided by ballast pockets.

7.10.3 Liferaft Equipment

Each liferaft is required to carry certain equipment that will provide assistance to the survivors and aid their location. The type of equipment supplied depends upon the area of operation of the vessel on which the liferaft is fitted. Standard equipment 'SOLAS A PACK' liferaft is required for vessels operating in an unlimited area of navigation whereas 'SOLAS B PACK' is for ships engaged on short international voyages. The equipment is given in Table 7.2.

Table 7.2 – Marine liferaft equipment								
Equipment	SOLAS 'A' PACK		SOLAS 'B' PACK		Equipment	SOLAS 'A' PACK		SOLAS 'B' PACK	
Rescue Quoit	✓	1	✓	1	Torch & Spare Battery/Bulbs	✓	2	✓	1
Buoyant Knife 1 for 12 persons 2 for > 12 persons	✓		✓		Radar Reflector or Radar Transponder Beacon (SART)	✓	1	✓	1
Buoyant Bailer Same as above	✓		✓		Daylight Signal Mirror	✓	1	✓	1
Sponges	✓	2	✓	2	Rescue Signal Card	✓	1	✓	1
Sea Anchors	✓	2	✓	2	Fishing Tackle	✓	1		
Buoyant Paddles	✓	2	✓	2	Food for each person	✓			
Tin Openers	✓	3			Water for each person	✓			
First Aid Outfit	✓	1	✓	1	Drinking Vessel	✓	1		
Whistle	✓	1	✓	1	6 Doses of Anti-seasickness each person	✓		✓	
Rocket Parachute Flares	✓	4	✓	2	Instructions Survival	✓	1	✓	1
Hand Flares	✓	6	✓	3	Repair Outfit	✓		✓	1
Buoyant Smoke Signals	✓	2	✓	1	Topping-up Pump	✓		✓	

7.10.4 Liferaft Container

The liferaft needs to be packed in a container that can withstand the harsh sea environment. It must provide sufficient buyoncy that can pull the entire length of the painter to activate the gas cylinder when the installation sinks without the liferaft being manually launched. In addition, the container should be marked with the following:

- maker's name, trade mark;
- serial number;
- approving authority name;
- liferaft carrying capacity e.g. 16 persons;
- SOLAS pack A or B;
- date of last service;
- length of painter;
- maximum permitted stowage height;
- instructions for launching.

7.11 Davit Launched Liferaft

Some installations may be fitted with SOLAS approved davits to launch Davit Launched Liferafts (DLR). The benefit of these liferafts is that all personnel can board at deck level and then it can be lowered into water using the davit.

These rafts are required to comply with normal design and construction standards but in addition, they must also be capable of withstanding impact against the installation at a velocity of 3.5 m/s and a drop into water from a height of at least 30 m. They must be provided with a means to keep them alongside the embarkation deck.

7.11.1 Launching Procedure – Davit Launched Liferaft

On command to 'Abandon ship' from the OIM or senior surviving officer, designated personnel prepare to launch liferafts following the procedure given below:

1. Expose lifting shackle.
2. Release wire brake and connect hook.
3. Remove container lashings.
4. Pull out and connect retaining line.
5. Pull out bowsing-in lines and secure to a strong point on deck.
6. Pull out release line – swing overboard keeping control of line.
7. Tend painter line – in event of release line failure this can be used as back-up.
8. Deploy liferaft(s) over the side.

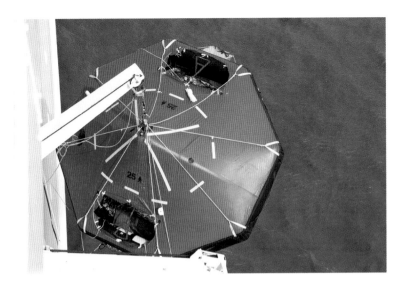

Figure 7.21 Davit launched liferaft

9.   Pull all excess painter line from the raft. When complete, give a sharp tug on the painter to inflate the liferaft.
10.   Pull liferaft close to ship and secure.
11.   Commence embarkation. Person in charge is to designate excess passengers into liferafts.
12.   Crew/passengers are to embark dry at deck level if possible. If necessary, secure raft to embarkation ladder using painter. Once designated personnel are aboard, person in charge of raft is to cut the painter, secure raft door and inflate the floor.
13.   Person in charge of raft can then slew liferaft clear of the installation's side using slewing handle then commence controlled descent to sea level by pulling on the lowering handle attached to the brake release wire.
14.   Person in charge is to hoist hook back to deck level and repeat launch procedure if more liferafts are required to be launched.
15.   These liferafts can then be towed away by rescue or lifeboats.
16.   Once clear of the danger area, deploy sea anchor and follow the instructions given in the 'Initial Actions to Be Taken in a Liferaft'.

7.12 Hydrostatic Release Unit (HRU)

HRUs are designed to release equipment such as liferafts, lifebuoys, EPIRBs or other similar equipment if the vessel to which they are fitted sinks without the equipment being released manually.

As the vessel goes under water, at a depth of not more than 4 m, the water pressure increases to a level high enough to activate the HRU knife that cuts the painter releasing the liferafts. For other equipment, the securing arrangement attached to a strap or bracket is cut releasing equipment such as EPIRB or lifebuoys.

Figure 7.22 HRU

It is advised that the person in charge of maintenance of LSA on-board installations should check the HRU expiry date, which is two years from the month of installation on-board. These units are disposable and cannot be serviced once expired.

The procedure below explains the securing arrangement for the HRU.

1. Attach the HRU with a shackle to a strong point on the deck or on the cradle.
2. Attach the liferaft lashing with a sliphook or similar to the white rope sling and firmly secure the liferaft.
3. Fit a shackle to the Red Weak Link connector and connect the painter line to the shackle.

In order to launch the liferaft manually, release the sliphook and throw the liferaft overboard. The liferaft is now attached to the vessel by the painter line via the Red Weak Link connector. Pull the painter line and inflate the liferaft so the survivors can board.

7.13 Aviation Liferafts

Rotary wing aircraft or helicopters use liferafts commonly referred to as helirafts which are stowed on-board the craft. These were first introduced into the North Sea in the 1980s. Generally, these can be deployed within about 60 seconds and are fully reversible unlike the marine liferafts which need to be righted if they capsize or inflate in a capsized position. When deployed, their canopy is stowed on the sides of the raft. The canopy can be pulled up by the occupants once they have boarded the raft. They can be supplied in different capacities depending upon the size of helicopter on which they are fitted. For example, helirafts are available in 7, 10, 14 or 18 person capacities. Due to helicopters' operational range from the nearest coast, it is unlikely that the survivors in the helirafts will be required to wait long for being rescued; however, helirafts are still required to be supplied with basic necessities for survival. Similarly, the principles and procedures that will apply to survive in a heliraft will be the same as for marine liferafts.

Standard equipment in aviation liferafts includes top-up pump, bailer, day and night hand flares, paddle, signal mirror, leak stoppers, sea anchor, safety knife, rescue line and quoit, whistle. Additional equipment may be supplied depending upon the area where a helicopter operates.

7.14 Initial Actions to Be Taken in a Liferaft

Initial actions in a liferaft are vital and if taken correctly, will increase the chances of survival significantly. These actions are based on four key areas and must be carried out in the order given here:

CUT – STREAM – CLOSE – MAINTAIN

7.14.1 CUT – The Painter

Once all survivors have boarded the liferaft (or lifeboat), then the person in charge should cut the painter to release the survival craft from the parent vessel. A knife, located close to the entrance to the liferaft, is provided for this purpose. The paddles can then be used to move the liferaft away from the vessel or installation keeping in mind the urgency for this action as the sinking vessel may pull the liferaft down with it.

7.14.2 STREAM – Drogue or sea anchor

When the liferaft has been moved to a safe distance away from the installation, it is important to maintain position in close proximity to the sunken vessel's last position. This is important because:

- The liferaft can pick other survivors from water.
- The other survival craft can congregate to assist each other and become a bigger target for SAR vessels.
- The sunken vessel would have transmitted its position prior to sinking. This is where the SAR is likely to begin. For offshore installations, the same rules apply even thought the last known position will not be an issue as it is likely to be in close proximity to the installation.

A sea anchor or drogue, permanently attached to the entrance to the liferaft, will enable the liferaft to reduce the rate of drift which otherwise is quite high for liferafts. Other benefits of drogue include:

- maintaining the liferaft entrances at an angle to the wind and waves;
- providing some resistance to liferaft capsize.

7.14.3 CLOSE – The Entrances

Extreme cold can cause hypothermia quite quickly. It is therefore recommended that the liferaft entrances are closed and liferaft floor inflated to preserve body heat and maintain the temperature within the liferaft. In hot weather, however, the entrances may only need to be closed when seawater sprays inside the liferaft. However, the drogue can be positioned appropriately to reduce the ingress of seawater and also to maintain the flow of air through the liferaft.

7.14.4 MAINTAIN – Condition, routine, lookout

Once the first three actions have been taken, then it is down to the survivors in the liferaft to maintain conditions within it. These include the physical condition of the liferaft with respect to repairing any damage and keeping it dry. In addition, survivors need to be kept busy; hence posting them on lookout for short periods will keep them busy and also assist in locating other survivors or SAR craft.

7.14.5 Secondary Actions to Be Taken in a Liferaft

Upon completion of the initial actions, some further actions need to be taken to ensure continued efforts for a successful survival. These include the following.

7.14.5.1 Anti-Seasickness Pills

All survivors, regardless of their resistance to seasickness, must be given a seasickness pill immediately. This is for two reasons:

1. After a short period of stay in the them, liferafts are known to cause motion sickness to even the best sailors.
2. Once one person becomes seasick, the smell of vomit will cause the others to follow.

If time permits, seasickness pills can be taken from the ship's medical chest and given prior to abandoning ship. Liferaft occupiers should also be issued with seasickness bags at the earliest opportunity.

7.14.5.2 Person In Charge of Liferaft

In order to maintain discipline on-board a liferaft, it is quite important to establish who is in charge of a liferaft. Normally, it will be the most senior ranked person from the deck department. If this is not possible, then the person who has the highest knowledge of operating survival craft and equipment should be given this opportunity through electing a suitable person or a person nominated in the muster list.

7.14.5.3 Injured Survivors

The first aid kit provided in the liferaft can be used to treat injured survivors. Priority should be given to the elderly, children and then depending upon the severity of injury.

7.14.5.4 Keep the Liferaft Dry

The liferaft will inevitably become wet and may continue to gather water in rough sea conditions. Occupants should be given sponges or bailers and assigned to keep it dry by bailing out as much water as they can.

7.14.5.5 Weather Conditions

It is important to maintain body temperature at all times. If survivors are cold and shivering, the easiest way to maintain temperature will be to huddle together and share body heat. If additional clothing and blankets are available, they can be used to further protect body heat. In case of the liferaft floor being too wet, lifejackets can be taken off and put on the floor with occupants sitting on them. TPAs (Thermal Protective Aids) provided in the liferaft can also be used for maintaining body temperature and conditions within the liferaft. If there are more than one survival craft in the area, survivors can transfer to one liferaft. Lifeboats provide better protection; hence consideration should be given to mustering all survivors in one lifeboat.

In hot weather, swimming should not be allowed to avoid loss of body fluid due to exhaustion. Furthermore, there is a danger of getting injuries from debris and other objects in the water. The liferaft floor can be deflated and position of the drogue adjusted to allow the flow of air through the liferaft.

7.14.5.6 Food and Water Rations

Consumption of food and especially water on-board a survival craft could be quite contentious, if not carried out carefully. As a general rule, no food or water should be given in the first 24 hours except for injured, sick or elderly persons and young children. Other exceptions to this rule could be the personnel who have been engaged in fighting fire or dealing with other emergencies prior to abandoning ship. Furthermore, the time at which the survivors had their last meal/water intake prior to abandoning ship should also be kept in mind as well as whether the survivors managed to bring additional supplies with them when they abandoned the vessel. Seawater should never be consumed as it will do more harm than good. Foods such as fish or other items containing protein increase thirst and therefore should be avoided unless a copious quantity of water is available.

7.14.5.7 Miscellaneous Actions

Once the liferaft routine has been established and initial and secondary actions have been taken, other actions to consider are:

1. Roll Call – Once all survival craft have congregated, a head count should be taken to establish the number of survivors, missing and injured persons. These steps are important for any further actions such as communication with SAR craft, search for other survivors, etc.
2. Sharp Objects – Any sharp objects brought by survivors should be removed to avoid damage to the liferaft.
3. Routine – The person in charge of the liferaft should take steps to establish a routine by assigning various tasks to all personnel e.g. keeping a lookout, bailing water out, looking after the injured, administering first aid, provision of food and water and repairing damage to the liferaft, etc. The lookouts should only be posted for brief durations by rotating them with others. The liferaft handbook should be given out for familiarisation with the procedures. All personnel should be mindful of the fact that their morale should always remain high. Consequently, they must all cooperate to enhance chances of survival.
4. Passing Urine – Urine retention could pose serious problems. Hence it is recommended that all personnel should try and pass urine within the first couple of hours of boarding a liferaft.
5. Location – Location devices such as SART, radar reflector, and EPIRB should be deployed at the earliest opportunity. The person in charge should endeavour to maintain the position of the survival craft.
6. Sharks – If the survivors find themselves in waters infested with sharks, they must keep all parts of their body covered and remain motionless if possible to avoid drawing a shark's attention. They should only move if they were to keep the shark in sight. Blood from wounds can quickly attract sharks, hence wounds should be covered. Any movement should be kept to minimum unless it is absolutely necessary. The best option will be to come out of the water.
7. Fire on Water – The survivors may have to remove their life-jackets to swim under water covered with fire. If holding breath underwater becomes difficult, then a clear space could be created by a sweeping movement of hands at the time of breaking through the water. At this time, the face should be covered to prevent injury from fire.

7.15 Pyrotechnics

The dictionary meaning of pyrotechnics is the art of making or displaying fireworks. However, in maritime survival context, pyrotechnics are the following equipment supplied in a SOLAS liferaft:

- Rocket Parachute Flares
- Hand Flares
- Buoyant Smoke Signals.

SOLAS lifeboats are supplied with six hand flares, four parachute rockets and two smoke floats. A brief description of other pyrotechnics is given below. Users must always read the instructions on the casing of each pyrotechnic prior to use.

7.15.1 Rocket Parachute Flares

Parachute flares can be fired to a height of 300 m with a descent rate of 5 m/second and then burn with a light of red colour for 40 seconds at an intensity of 30,000 candelas. Care should be taken when firing parachute flares near an approaching aircraft. If fired directly toward the aircraft, it will be seen as ammunition that can potentially cause damage and hence it will steer away.

7.15.2 Hand Flares

Hand flares burn for a period of 1 minute at an intensity of 15,000 candelas. When igniting hand flares, care should be taken to hold it away from the windward direction holding it from the position indicated on the flare to avoid burning the hands.

7.15.3 Orange Smoke Floats

These burn for 2–4 minutes and emit orange coloured smoke. These should be used to the leeward side of the survival craft to pinpoint the survival craft's position for SAR vessels.

7.16 The Rescue Operation

Generally, emergencies take place at night and in extreme weather conditions. Whilst the distress signal may have been transmitted as soon as the installation's emergency situation became unmanageable, the rescue operation may still take some time to commence and the SAR vessel to find the survivors in the survival craft. By the time the SAR vessels arrive at the scene, the weather conditions may not have changed, making the rescue operation extremely hazardous. To complicate matters further, the survivors' injuries may have been aggravated during their time on the survival craft, they may have been weakened due to thirst and hunger and may become 'desperate' on seeing the SAR craft. It is of utmost importance that the person in charge of the survival craft takes control of the situation even during the rescue operation. For example, when personnel have to board another vessel, a safety harness should be used to

Figure 7.23 Helicopter rescue operation – winchman being pulled towards the vessel

prevent them falling into the water even if a boarding ladder was deployed to climb on-board. Another technique that will be used by rescue helicopters is known as the Hi-Line technique. This is described in the next subsection.

7.16.1 Helicopters

SAR helicopters are equipped with VHF 16 which can be used for communication with the survival craft.

When the SAR helicopter has located the survival craft, the survivors need to take the following actions in preparation of the Hi-Line rescue.

1. All loose items in the survival craft should be secured to avoid being blown away with the helicopter downdraft from the rotors.
2. Any items erected at high position such as a radar reflector or SART should be taken down.
3. Other electronic equipment should be switched off to avoid interference with helicopter's navigation systems.
4. The sea anchor should be deployed to reduce the rate of drift in the survival craft.
5. The floor of the liferaft should be deflated to provide more resistance to movement in water.
6. Secure the canopy of the liferaft to provide as large a clear area as is possible.

Once the above actions have been taken, the instructions of the helicopter crew should be complied with fully. If VHF radio is available, the helicopter pilot will be able to communicate with the survival craft using the given channel.

Figure 7.24 Helicopter rescue operation

In certain weather conditions it may not be possible to winch the helicopter winchman or the strop (rescue harness) from a position directly above the survival craft. Under such circumstances a weighted rope extension to the winch wire may be lowered. This extension is known as a Hi-Line and is connected via a weak link to the aircraft's winch hook. Once the weighted line is placed in the survival craft, one person must handle the line. They should take up the slack on the Hi-Line and haul in ONLY when instructed to do so by the helicopter crew by radio message or hand signal. The Hi-Line must NOT be secured to any part of the vessel and instead can be coiled into a bucket or similar container clear of obstructions.

If the helicopter has to break away during the operation the line must be paid out or, if necessary, released completely ensuring that the line passes clear outboard. As the Hi-Line is paid out, the helicopter will move to one side of the vessel and descend. Normally the winchman will be winched down; the person on survival craft will continue to take in the slack. As the winchman or strop approaches the vessel, the earthing lead or hook must make contact with the water to discharge the static electricity before any person on survival craft makes contact with the wire.

Upon passing the helistrop around the person to be rescued, the person in charge in the survival craft should signal to the helicopter when it can start to winch in the wire (this task may not need to be carried out if the winchman was sent down from the helicopter). As this occurs a crew member should pay out the Hi-Line, maintaining sufficient firmness to prevent any swing.[4]

7.16.2 Fast Rescue Craft (FRC)

Fast rescue craft are carried on most ERRVs, offshore installations and vessels to provide fast response for any rescue operations. These may also be fitted on many offshore fixed installations or MODUs. When FRC are available, the coxswain must have undergone approved training such as ERRV Crew Fast Rescue Craft Coxswain or Boatman or STCW Proficiency in Fast Rescue Boats. Generally a FRC is of RIB (Rigid Inflatable Boat) construction but many are now rigid structures made of either GRP or aluminium hulls. Normal rescue boats may also be of the same construction but if categorised as FRC, they must comply with the SOLAS fast rescue boat requirements summarised below:

1. hull length including inflated structure should be not less than 6 m and not more than 8.5 m;
2. provided with sufficient fuel to manoeuvre the boat for at least 4 hours at a speed of 20 knots in calm water and carrying 3 crew members and at a speed of 8 knots carrying full complement of persons and equipment;

Table 7.3 – Fast rescue craft equipment

Equipment	FRC		Rescue boat		Equipment	FRC		Rescue boat	
Boat Hook	✓	*	✓	*	Radar Reflector	✓	1	✓	1
Oars/Paddles	✓	*	✓	*	TPA	✓	2**	✓	2**
Bailer	✓	1	✓	1	Fire Extinguisher	✓	1	✓	1
Compass	✓	1	✓	1	Bucket (Rigid Rescue boat)	✓	1	✓	1
Sea Anchor	✓	1	✓	1	Knife/Hatchet	✓	1	✓	1
Painter	✓	1	✓	1	Safety Knife (RIB)	✓	1	✓	1
Buoyant Line	✓	1	✓	1	Sponge (RIB)	✓	2	✓	2
Torch with Batteries	✓	1	✓	1	Bellows/pump (RIB)	✓	1	✓	1
Whistle	✓	1	✓	1	Repair Kit (RIB)	✓	1	✓	1
First Aid Kit	✓	1	✓	1	Safety Boat Hook (RIB)	✓	1	✓	1
Rescue Quoit	✓	2	✓	2	Search Light	✓	1	✓	1

* Sufficient quantity
** or 10% of the boat capacity

3. shall be self-righting or capable of being righted and self-bailing or capable of being cleared of water; when righted after a capsize, the engines should be capable of being restarted;
4. shall have means to switch the engine off in case of accidental capsize;
5. shall be fitted with a watertight VHF radio;
6. shall be provided with the equipment as given in Table 7.3.

7.17 Global Maritime Distress and Safety System (GMDSS)

The Global Maritime Distress and Safety System (GMDSS) has been in use since 1992 but was fully implemented on 1 February 1999. The system's major components consist of ship based radio systems, satellites connected with shore based communication systems to transmit alerts related to incidents at sea. These alerts can also be received through satellite communications systems by other ships or shore based establishments.

Mandatory GMDSS requirements apply to ships of 300 gt and above and all passenger ships engaged in international voyages. The salient feature of GMDSS is that when a distress message is transmitted, the SAR operation will commence. This includes times when the ship's personnel are unable to transmit the alert manually.

Whilst primarily intended for ships at sea, the same arrangements apply to offshore installations.

Under GMDSS requirements, the world's seas and oceans are divided into four areas on the basis of distance of vessels from the coast. This division also determines the type of equipment these ships are required to carry. These sea areas are:

- **A1** – an area within radiotelephone coverage (typically 20–30 miles) of at least one VHF coast station in which continuous DSC (Digital Selective Calling) alerting is available.
- **A2** – excluding Area A1 but within radiotelephone coverage (typically 150–250 miles) of an MF coast station in which continuous DSC alerting is available.
- **A3** – excluding sea areas A1 and A2, but within INMARSAT geostationary satellite coverage, in which 'continuous alerting' is available. This coverage is between 76 degrees north and 76 degrees south latitudes.
- **A4** – any area excluding sea areas A1, A2 and A3.

All ships to which GMDSS requirements apply, irrespective of sea area, shall be fitted with the following:

- VHF and VHF/DSC radio installation;
- SART operating in 9 GHz band or AIS SART;
- NAVTEX receiver and INMARSAT C EGC (Enhanced Group Calling) receiver for areas where NAVTEX service is not provided;
- Satellite Emergency Position Indicating Radio Beacon (EPIRB).

Ships navigating only in coastal areas (A1) do not require a satellite communication system but they may choose to use it.

7.17.1 EPIRB

EPIRB is an acronym for Emergency Position Indicating Radio Beacon. The EPIRB is used to transmit a distress message to search and rescue services and other vessels in the vicinity via the satellite based communication system. EPIRBs are fitted with float free arrangements which allow release of the EPIRB and its activation if the vessel sinks without someone manually activating it.

7.17.1.1 VHF EPIRB

Vessels sailing exclusively in sea area A1 can carry a VHF EPIRB instead of a satellite EPIRB. VHF EPIRBs use VHF channel 70 to transmit the nature of distress and if there is a built-in GPS, then the position information is also transmitted. The locating signal may in addition be provided by means of a 9 GHz radar transponder.

7.17.1.2 INMARSAT EPIRB

This EPIRB works on 1.5–1.6 GHz frequency using INMARSAT's geostationary satellites. Since geostationary satellites are used, alerts are transmitted nearly instantly to a rescue coordination centre associated with the INMARSAT coastal earth station receiving the alert.

7.17.1.3 COSPAS SARSAT EPIRB

This EPIRB operates on frequency 406/121.5 MHz. COSPAS SARSAT polar orbiting satellites provide global coverage. The satellite determines the position of EPIRB to within 5 km (3 miles) by using the satellite's own position and Doppler shift. This information is then transmitted via an earth station to the nearest rescue coordination centre (RCC). This system offers a global coverage since the satellites are in a polar orbit.

7.17.2 SART

SART stands for Search and Rescue Radar Transponder. There are two types of SART given below.

7.17.2.1 Radar SART

Radar SART operates on a frequency of 9.2–9.5 GHz and is a passive transmitter i.e. when switched on, it will remain in standby mode but when interrogated by radar waves, it will transmit signals that will show on the radar screen of a 3 cm radar. Typically, it can be detected to a range of up to 12 miles where range and bearing can be used to locate position of SART and therefore the survival craft.

7.17.2.2 AIS SART

AIS (Automatic Identification System) SART can be used in the same way as Radar SART but in this case, the signal transmitted from AIS SART will show on an AIS receiver on-board ship. This type of SART is considered more reliable due to higher positional accuracy transmitted from a built-in GNSS receiver. However, the receiving ship must have its AIS on to receive the transmitted signal.

Since 1 January 2010, IMO/SOLAS allows the use of approved AIS SART as an alternative to the Radar SART. Its detection range is:

- up to 10 nm from a SOLAS ship with AIS Class A transponder;
- up to 40 nm from a SAR helicopter (at an altitude 1000 feet);
- up to 130 nm from a SAR aircraft (at an altitude 20,000 feet).

7.17.3 Emergency Locator Beacons

Emergency locator beacons transmit a signal that can be tracked via satellite or through radio direction finding facilities. These are now commonly used to aid the location of small vessels, aircraft or persons in distress. Some beacons can transmit signals to satellites such as EPIRB, others can transmit signals to radar such as SART and some can transmit signals to radio direction finding equipment such as personal location beacons. A description of other types of beacons is given in the next section.

7.17.4 Aircraft Beacons

The rotary wing aircraft used in the offshore industry are fitted with Emergency Location Transmitters (ELTs) which can be activated manually by the pilot/co-pilot or automatically when the aircraft submerges under water. The ELT beacon has a built-in GPS which allows it to transmit a distress message along with its position on 406 MHz to the satellite. Another transmitter within the same beacon transmits signals on 121.5 MHz for homing through the use of direction finders.

Helirafts are also fitted with an ELT which can be activated manually by the occupants.

7.17.5 Personal Locator Beacon (PLB)

PLB is a Maritime Survivor Locating Device (MSLD) which can either operate on single or dual frequencies i.e. 406 MHz for satellite signals and 121.5 MHz for homing signals. Many units operate

Figure 7.25 PLB with GPS (Source: www.crew-safe.co.uk/)

only on one frequency i.e. 121.5 MHz to assist in locating survivors by using a homing device. These are particularly useful for MOB situations to aid quick location. They are of rugged float free construction, bright orange in colour and fitted with a flashing light. Helicopter pilots and passengers (in the UK) are supplied with a PLB that usually fits into a lifejacket.

PLBs can be activated automatically upon submersion in water or manually by switching them on. Their battery life is a minimum of 12 hours in transmission mode. Some models are designed to transmit signals on VHF DSC frequency as well which will be received by all ships equipped with DSC receivers. For VHF signals, the range is usually up to 10 miles due to less height above the surface of water.

In 2009, PLBs were withdrawn from use in the UK North Sea sector due to known signal interference issues with the aircraft locator beacons in a helicopter ditching on 18 February 2009 where PLBs caused the aircraft beacons to shut down. However, after extensive research, new beacons which do not interfere with the aircraft beacons were introduced in February 2010 and are currently used widely.

Notes

1 HSE (2001) Inspecting and Auditing the Management of Emergency Response. Offshore Technology Report 2001/091. Available at www.hse.gov.uk/research/otopdf/2001/oto01091.pdf [Accessed 07.12.14].
2 EPOL (2014) Integrated Offshore Emergency Response. Available at www.epolgroup.co.uk/files/4113/9566/4365/Integrated_Offshore_Emergency_Response__Version_1.2.pdf [Accessed 07.12.14].
3 MCA (2009) MGN 388 – Fitting of Fall Prevention Devices. Maritime and Coastguard Agency, Southampton, UK.
4 MCA (2000) MGN 161 (M + F) – Search and Rescue Helicopter Hi-Line Technique, Maritime and Coastguard Agency, United Kingdom.

8

First Aid in Survival Situations

In any situation where personnel may be injured, the importance of first aid cannot be over-emphasised. In essence, first aid encapsulates four objectives:

1. Save life.
2. Ease suffering.
3. Prevent further harm.
4. Support recovery.

Whilst all offshore workers will have a basic knowledge of first aid techniques, specialist medical support is provided through the offshore medic who has access to medication and facilities to treat the injured and look after others who require initial medical treatment. In addition to the medic, a number of other workers would have undertaken specific first aid training required for nominated first aiders or first responders.

Full knowledge of first aid provision is beyond the scope of this book as it only addresses the basic principles in survival situations as required for BOSIET, FOET and TEMPSC courses. The description here is only to act as a reminder for the knowledge gained on the full first aid courses and to provide emergency first aid in survival situations.

8.1 Prioritise First Aid Actions

For correct sequence of first aid actions, DR ABC can be used as an aide-memoire as given in Figure 8.1:

D Danger – Ensure neither you nor the casualty is in any danger.
R Response from the casualty by trying to talk to them.
A Airway – Whether it is open or not.
B Breathing – Is the casualty breathing?
C Circulation – Can be checked by checking casualty's pulse rate. Quality of circulation can also be assessed.

Figure 8.1 First Aid Priorities

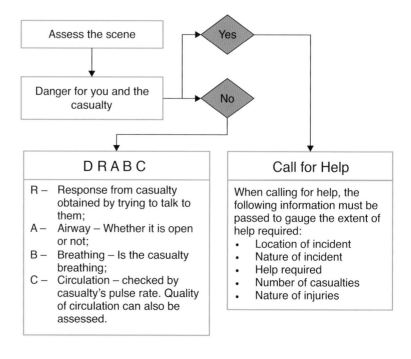

8.2 Cardiopulmonary Resuscitation (CPR)

CPR is a first aid technique that can be used for casualties who are not breathing properly, have stopped breathing or their heart has stopped.

An early action can save someone's life, therefore whilst the first person on the scene is waiting for the medic to arrive, they can commence CPR as soon as possible. The objective for CPR is to ensure exchange of CO_2 and oxygen continues in the lungs to avoid causing 'brain' death due to lack of oxygen. In addition, CPR will also support flow of blood through the body due to the heart being compressed between the chest and backbone. There are two ways in which CPR can be administered.

8.2.1 Hands Only CPR

At times it may not be convenient to give a mouth-to-mouth CPR due to either the first aider being uncomfortable with any fluids in the mouth or due to lack of training. In such cases, instead of doing nothing, hands only CPR can be carried out through applying compressions on the casualty's chest. The sequence for hands only CPR is as follows:

* Check casualty's airway to ensure it is not blocked. This can be done by keeping one hand on the casualty's forehead and tilting

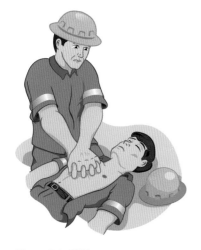

Figure 8.2 CPR

their head back by lifting the chin up with two fingers. Whilst doing this, you can also observe the rise/fall of the casualty's chest to see signs of breathing.

- Place heel of the hand in the centre of the casualty's chest on the breastbone.
- Place second hand on top of first hand interlocking fingers of both hands as shown in Figure 8.2.
- Keep shoulders in line with and right above hands.
- Press the chest down by 5–6 cm by using the weight of the body. Approximately 100–120 compressions per minute should be carried out until the medic arrives.

8.2.2 CPR with Rescue Breaths

Personnel who have been trained in this technique can easily follow the procedure given below to increase the chances of the casualty's survival:

- Follow the procedure for hands only CPR i.e. heel of one hand in the centre of casualty's chest with the second hand on top of first and fingers interlocked. Push the chest down by 5–6 cm at a rate of 100 compressions per minute.
- After 30 compressions, deliver two mouth-to-mouth breaths. In order to do this, tilt the casualty's head by lifting the chin up with two fingers. Pinch the nostrils with thumb and one finger and then seal lips with casualty's lips. If a face shield is available, then it should be used instead of direct mouth to mouth. This can be done by a second person available. Whilst breathing into the casualty, eyes should be kept on the casualty's chest to see it rising. Once you have breathed into the casualty, move your mouth away and see the chest falling and then give the second rescue breath.
- After delivering the second breath, give another 30 compressions.
- The above cycle should be continued until the casualty recovers or the medic arrives.
- If the casualty recovers, then place them in the recovery position described later in this section.
- If you can't deliver breath into the casualty, then check the airway again which may have been blocked.
- For a drowning casualty: before commencing compressions, the casualty should be given 5 rescue breaths followed by 30 compressions and 2 rescue breaths. This sequence of 30 compressions and 2 rescue breaths should then be continued.
- The above procedure can be used for both adults and children.

Figure 8.3 CPR Priorities

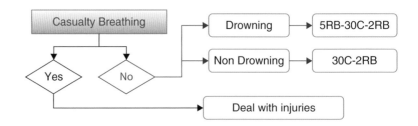

8.2.3 Recovery Position

The Resuscitation Council (UK) recommends the following sequence of actions to place a victim in the recovery position:

- Sit beside the casualty with your knees on the deck.
- Straighten the casualty's legs with the arm away from you alongside the body of the casualty.
- The arm closest to you should be bent in a 'L' shape at a right angle to the casualty's body and the palm parallel to their head.
- Move the casualty's arm placed away from you over the chest as if the casualty were trying to hold their neck.
- Hold the furthest leg from the knee and raise it up whilst ensuring the casualty's hand is kept in position close to their neck.
- Maintaining this position, roll the casualty over towards you by pulling on the furthest leg.
- Once rolled over, adjust the position of the casualty with one hand under the cheek with knee bent at right angle to the body. The head should be tilted back to ensure the airway remains open.
- Continue to monitor the casualty once in the recovery position.

Figure 8.4 Recovery position

8.3 Dealing with Injuries

8.3.1 Management of Shock

Shock is a medical condition associated with a fall in blood pressure often resulting from severe pain, injury or medical conditions such as a heart attack and leading to reduced vital activities of the body. If the casualty is conscious, the first person on the scene should reassure them and then follow the guidance given in the flow chart in Figure 8.5. If not treated properly, it could lead to immediate death.

8.3.2 Management of Bleeding

Bleeding may be caused due to injuries on-board or during abandoning the installation. The two consequences of bleeding are shock

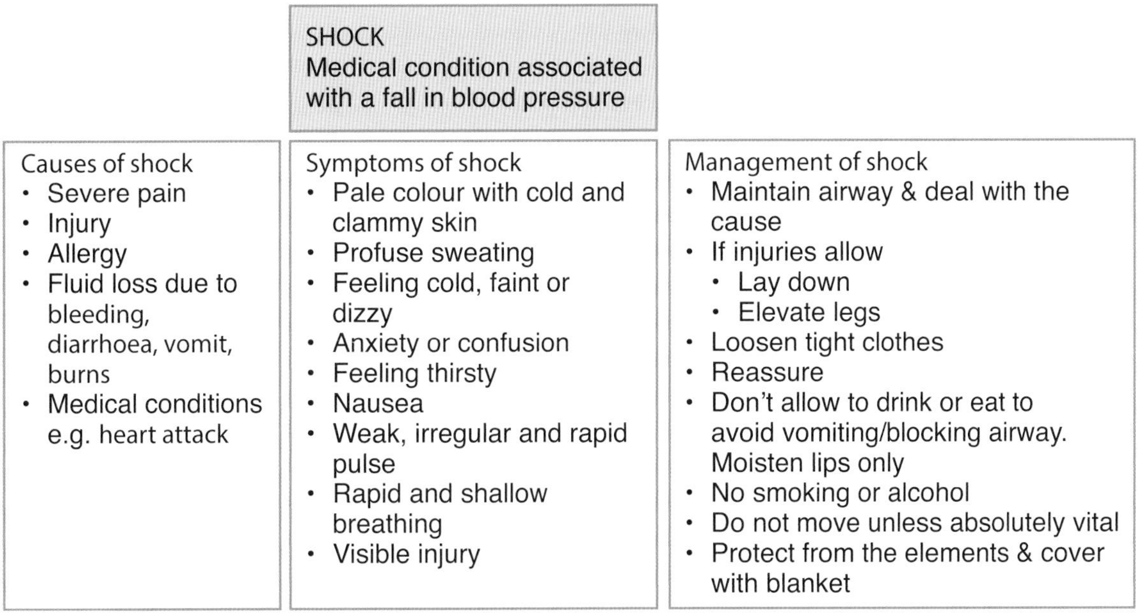

SHOCK Medical condition associated with a fall in blood pressure		

Causes of shock
• Severe pain
• Injury
• Allergy
• Fluid loss due to bleeding, diarrhoea, vomit, burns
• Medical conditions e.g. heart attack
	Symptoms of shock
	• Pale colour with cold and clammy skin
	• Profuse sweating
	• Feeling cold, faint or dizzy
	• Anxiety or confusion
	• Feeling thirsty
	• Nausea
	• Weak, irregular and rapid pulse
	• Rapid and shallow breathing
	• Visible injury
		Management of shock
		• Maintain airway & deal with the cause
		• If injuries allow
		  • Lay down
		  • Elevate legs
		• Loosen tight clothes
		• Reassure
		• Don't allow to drink or eat to avoid vomiting/blocking airway. Moisten lips only
		• No smoking or alcohol
		• Do not move unless absolutely vital
		• Protect from the elements & cover with blanket

and loss of blood. Shock should be treated as described earlier alongside the treatment of the bleeding as given here.

Figure 8.5 Shock management

There are three types of bleeding:

1. Arterial Bleeding: when the arteries are injured through deep cuts, it causes bright red blood to spurt. The flow of blood is steady and in copious quantities. The blood will appear to spurt out with each heartbeat. In such bleeding, direct and firm pressure on the wound needs to be applied to stop bleeding. If delayed, blood loss can lead to death within minutes;
2. Venous bleeding: injuries to veins causing dark red blood to flow at steady but slow rates. In any such bleeding, direct pressure should be applied to the affected vein.
3. Capillary bleeding: minor scrapes or cuts can damage body capillaries leading to a slow and small quantity blood flow which will be controlled by the body's natural clotting mechanism in a matter of minutes.

Other precautions for bleeding are given in the flow chart in Figure 8.6.

8.3.3 Management of Burns

A burn is an injury caused by exposure of body parts to heat resulting from fire or chemicals. In survival situations, there is a very limited amount of help available, therefore the guidance in the flow chart in Figure 8.7 should be followed.

Figure 8.6 Management of bleeding

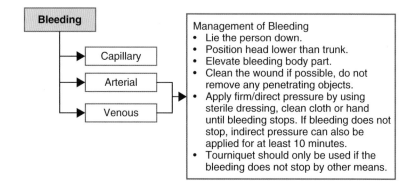

Figure 8.7 Management of burns

Burn
an injury caused by exposure to heat, flame or chemicals

Causes	Severity	Management
• Dry heat • Wet heat • Electrical • Chemical • Radiation • Friction	Measured by depth and area of burn **DEPTH** • **Superficial** – *red, tender, no blisters* • **Partial** – *red, tender, blisters* • **Deep** – *all layers, appearance varies (e.g. white, charred)* **AREA** The palm of the casualty is considered to be approximately 1% of their body surface. Its size can be used as a template to measure the total body area burnt.	• Remove from source of heat • *Cool for at least 10 minutes* • Remove constrictions • Cover with non-fluffy dressing

8.3.4 Cold Injuries

Cold injuries can be caused by exposure of body parts to extremely low temperatures that may result in a localised injury or cooling of the entire body leading to hypothermia. Severity of the injury will depend upon the temperature, physical status of individual's health, protective clothing and duration of exposure. Cold injuries are divided into two categories:

• Non-freezing cold injuries: caused by exposure to wet conditions combined with low temperatures. For example, immersion foot where the foot may become black due to injury. If appropriate first aid precautions as given in the flow chart in Figure 8.8 are not taken, then within 24–48 hours, injury could lead to permanent tissue damage.

• Freezing cold injuries: these injuries occur due to body parts such as hands, feet, nose, lips and ears being exposed to freezing temperatures. The mildest of this type of injury is called 'frostnip' which will cause numbness and severe pain in the area affected. However, upon warming, the effect of injury is reversible. If frostnip is not controlled through rewarming, it can develop

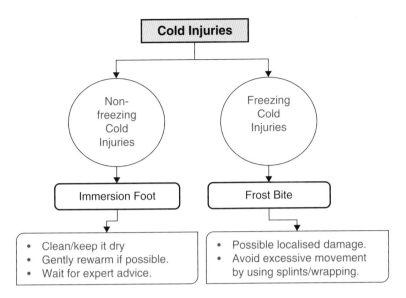

Figure 8.8 Management of cold injuries

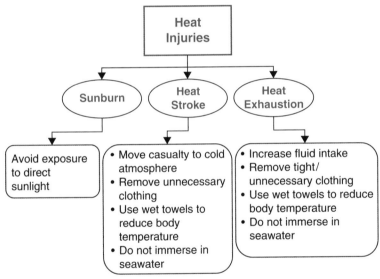

Figure 8.9 Management of heat injuries

into frostbite where the tissues will freeze and eventually die if not treated immediately.

8.3.5 Heat Injuries

Remaining in a survival craft in an extremely hot climate can lead to heat injuries such as sunburn, heat stroke or heat exhaustion. Sunburn is caused by exposure of body parts to direct sunlight whereas the other two injuries could be a result of the loss of body fluids resulting in dehydration combined with excessive salt depletion. For first aid actions, see Figure 8.9.

Figure 8.10 Ailments

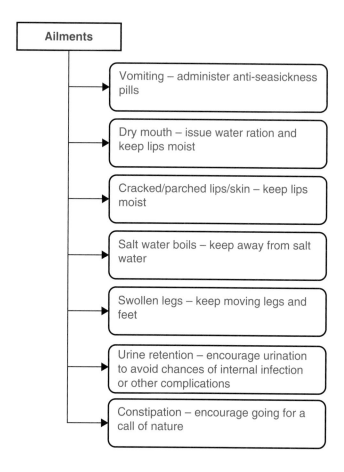

8.4 Ailments

Various ailments can affect survivability in different ways. The flow chart in Figure 8.10 shows some of the most common ailments and some advice to remedy them.

8.5 Cold Water Immersion and Hypothermia

Evaporation is a cooling process which is evident during sweating, i.e. the evaporation of sweat cools the body providing evidence of this basic principle. Even when the weather is warm, a sudden immersion into seawater could cause a reduction in body temperature. The same principle applies whenever the water temperature is lower than the body temperature i.e. the water will take away heat from the body and reduce its temperature.

The severity of impact will depend upon the temperature of water, but in any case, the stages of impact as given in Figure 8.11 need to be considered.

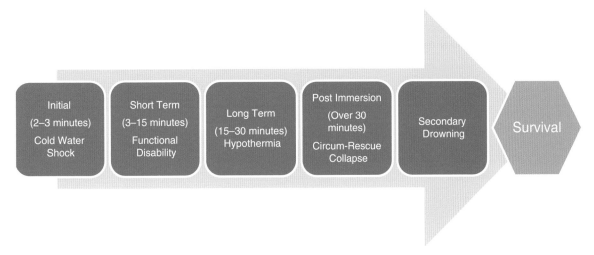

Figure 8.11 Hypothermia stages

8.5.1 Initial Impact

A sudden entry into cold water may cause an unexpected gasp commonly referred to as 'cold water shock' that will allow water to enter lungs. If this happens, hyperventilation is likely leading to drowning almost instantaneously. It will be difficult to hold breath during this time which can be combined with 'tachycardia' i.e. rapid heartbeat. This can happen within the initial 2–3 minutes. As a precaution and if circumstances permit, cover the face (both mouth and nose) with a hand upon impact with the water and subsequently keep the head out of the water. Due to a sudden change in body temperature, chances of cramps are also increased which may lead to other complications.

8.5.2 Short-Term Impact

Due to the sudden loss of body heat combined with exhaustion, drowning may occur if the person survives the initial impact. Even expert swimmers may find themselves in a situation where their arm and leg movements to remain afloat will be affected resulting in inhaling water and therefore drowning. A lifejacket, preferably fitted with a face visor, will prove to be very effective during this phase which usually occurs between 3 and 15 minutes.

8.5.3 Long-Term Impact

If the person continues to remain in water after the short-term impact phase, progressive cooling of the body continues, leading to hypothermia. Initial signs of hypothermia can be seen in this phase which lasts up to about 30 minutes from initial immersion in water.

Table 8.1 Stages of hypothermia			
Hypothermia Stage	**Body Temperature**	**Duration in Water**	**Impact**
Mild	35°C	> 30 minutes	Shivering, numbness, slow response, slurred speech, blurred vision, irrational behaviour
Moderate	32°C	30–120 minutes	Cessation of shivering, abnormal heartbeat, unconscious, possible drowning without lifejacket
Severe	28°C	90–180 minutes	Vital signs reduced or absent, ventricular fibrillation or cardiac arrest
Death	24°C	Over 180 minutes	

8.5.3.1 Management of Hypothermia

Restore Body Temperature Slowly	• Replace wet clothing with dry clothing. • Warm with extra layers of clothing e.g. blankets etc. • Monitor temperature if possible. • Use TPAs or spare lifejackets if nothing else available. • Huddle together to share body heat whilst gently handling the casualty to avoid exertion which can increase heartbeat leading to blood flow problems. • When in SAR vessel, lay the casualty down horizontally with legs slightly raised. Avoid using hot water bottles but use extra blankets. Do not rub/massage limbs to warm them up. Alert casualty may be given a warm shower/bath but under supervision at all times. Warm drink may be given to alert casualty. No alcohol should be allowed. • If casualty is not breathing, commence CPR. In all cases, seek further medical advice.
Begin CPR (if required)	• Check casualty's response. • Put heel of your hand on the centre of casualty's chest and other hand on your first hand lacing fingers together. • Keep arms straight with shoulders above hands. • Push down on your hands compressing casualty's chest to about 5–6cm. Compress the chest to a depth of 5–6 cm and at a rate of 100–120 per min. • Give each rescue breath for 1 second. • Do not stop to check the victim or discontinue CPR unless the victim starts to show signs of regaining consciousness, such as coughing, opening their eyes, speaking, or moving purposefully AND starts to breathe normally. • Ratio of 30 compressions to 2 breaths should be maintained.
Give Warm Fluids	• Once the casualty is breathing and alert, give warm fluids avoiding alcohol and caffeine. • Do not leave the casualty unattended. • Continue to monitor casualty's body temperature.
Maintain Body Temperature Up	• Even when the casualty is alert, do not rush to remove wet clothing. Maintain the body temperature by keeping them wrapped in blanket.
Follow up	• Keep the casualty under observation until expert advice is available and further action is recommended according to the condition of the casualty.

Figure 8.12 Hypothermia management

8.5.4 Post Immersion Impact and Hypothermia

After remaining in water for over 30 minutes, onset of hypothermia is extremely likely. The body would have lost a significant amount of heat, literally cooling it down to 'unconsciousness'. Blood flow to the extremities of the body reduces in an effort to preserve heat to protect blood flow to the vital organs – a phenomena known as vasoconstriction. In such a situation, muscles wouldn't be able to perform as required resulting in loss of meaningful movement of arms and legs. In such circumstances, if the head goes under water as a result of unconsciousness, drowning is considered imminent. If, however, the head remains above water due to, for example, a lifejacket, then cardiac arrest may occur due to loss of body heat.

Hypothermia is a condition of the human body which results when the body loses more heat (due to shivering and contact with water) than it produces. Hypothermia can be considered in three stages depending upon the body temperature.

8.5.5 Secondary Drowning

Also known as 'delayed' or 'dry drowning', 'secondary drowning' is a phenomena which may occur sometime after a person has been rescued from water. The person may appear fit and well but the water that was inhaled prior to being rescued is absorbed into the lungs and affects the exchange of oxygen and carbon dioxide in the lungs. Generally, this is evident from symptoms such as gasping, coughing, wheezing or lethargy.

Generally, a survivor suspected of such a situation should be transferred to hospital if possible otherwise they should be kept under observation and radio medical advice should be sought.

8.5.6 Circum-rescue Collapse

Circum-rescue collapse is a phenomenon that may occur just before, during or shortly after rescue. Causes of this include a reduction in the body's ability to maintain temperature resulting from improper heart function, loss of blood pressure due to post-immersion problems or even the minor 'rescue' factors such as:

*   The 'news' of being rescued causes an increase in the adrenaline level in the body and thereby slows its response in maintaining the blood pressure resulting in a cardiac arrest immediately before being rescued.
*   The heart is already 'cold' due to overall loss of body heat and cannot pump cold blood effectively to maintain blood pressure.
*   The position in which a person is lifted from water plays a major role in the chances of survival after being rescued from water. Hydrostatic squeeze on the 'immersed' parts of the body reduces

the blood flow to those parts. When the body is lifted vertically, the blood rushes to the lower body causing a sudden reduction of blood supply to the heart itself and the brain, which may cause fainting or cardiac arrest even leading to death.

As a result of the above, both the survivors and rescuers should keep in mind that the horizontal lift should be preferred for those who have been in water for some time.

Bibliography

St John Ambulance www.sja.org.uk
NHS Choices www.nhs.uk
British Red Cross www.redcross.org.uk/en/What-we-do/First-aid
Resuscitation Council (UK) www.resus.org.uk

9

Helicopter Safety and Escape

Helicopters are considered workhorses of the offshore oil and gas industry because of their ability to transport personnel to and from the offshore installations, but accidents continue to happen despite efforts by operators and other industry stakeholders. A synopsis of the major helicopter incidents that took place in the UK oil and gas sector are given in Table 9.1.

It is estimated that offshore industry in the EU member states use around 230 helicopters of ten different types from three manufacturers. Whilst consideration of the design features of helicopters is beyond the scope of this book, various components relevant to passengers travelling offshore are discussed with a training requirement for offshore workers.

Expansion of the offshore oil and gas industry in the late 1960s led to an increased use of helicopters as a means of transportation. However, this was unfortunately combined with a drop in safety rating for this mode of travel as described earlier. It was not until 1986 that all offshore workers were required

Figure 9.1 Helicopter landing on an offshore rig

Table 9.1 – History of major helicopter accidents in the oil and gas industry[1]			
Accident Date	**Helicopter Type**	**Location**	**Nature of Accident**
21/04/1976	Sikorsky S-58 ET	Forties Field	Fall during forced landing, 1 death out of 10 persons on-board
17/04/1978	Aerospatiale AS330J	North Sea	Crew lost control during takeoff, tail rotor struck ground. No injuries, 12 persons on-board
12/08/1981	Bell 212	North Sea	Descent into sea due to crew losing control 1 death out of 14 persons on-board
13/08/1981	Westland Wessex Mk 60	North Sea	Crash into sea resulting from loss of engine power. All 13 persons on-board died
20/11/1984	Bell 212	North Sea	Air dived into sea after a loud bang. All 2 persons on-board died
06/11/1986	Boeing BV234LR	North Sea	Crash into sea, 45 deaths out of 47 persons on-board
25/07/1990	Sikorsky S-61N	North Sea	Crash into sea after tail rotor striking the platform, All 6 persons on-board died
14/03/1992	Aerospatiale AS332 L1v	North Sea	Crash into sea, 11 deaths out of 77 persons on-board
18/04/1992	Sikorsky S-76	North Sea	HLO killed by rotor blade. No fatalities on helicopter
22/09/1992	Aerospatiale SA365 N	North Sea	Helideck crew killed by rotor blade. No fatalities on helicopter
16/07/2002	Sikorsky S-76A	North Sea	Crash into sea due to rotor failure, All 11 persons on-board died
27/12/2006	Eurocopter AS365 N2	Irish Sea	Crash into sea, All 11 persons on-board died
01/04/2009	Eurocopter AS332 L2	North Sea	Crash into sea, 4 deaths out of 18 persons on-board
23/08/2013	AS332 L2 Super Puma	North Sea	Crash into sea due to gearbox failure, All 16 persons on-board died
12/12/2013	Sikorsky S-76	Bintulu, Malaysia	Ditching into sea. 6 persons on-board. No fatalities/injuries

to undertake Sea Survival training that included Helicopter Underwater Escape Training (HUET).

Today, offshore operators have mandated this training for all offshore travellers in most of the world but recent accidents show room for further improvement. Whilst the relevant sectors of the industry are looking at improvements in overall safety when travelling by helicopter, training issues need to be recognised and rectified by all concerned. The industry has generally improved safety during escape from a ditched helicopter through design of the seats, window release mechanisms, improved SAR capability as well as improvements in training.

The main issue that the survivors have to deal with is disorientation and therefore lack of ability to locate and use the escape windows. However, other significant factors such as daytime versus night time survival, effect of cold temperature of water causing hypothermia, ability to use survival equipment, possibility of injuries during escape from the helicopter are some of the areas that offshore workers need to be aware of to increase their chances of survival.

It is important here to distinguish between the two terms used in the context of HUET training and/or helicopter emergencies. These are 'ditching' and 'crashing' with the former being controlled or semi-controlled landing on surface of water whereas the latter is an accidental fall into water.

In the case of ditching, the time available to put training into practice and survive is considered to be more than in a crashing scenario. Research indicates that about 30–50 seconds are required to exit from a ditched helicopter under normal conditions i.e. upright helicopter landed on the surface of water in calm seas. However, this figure does not include other eventualities such as the impact of sea, waves, wind and any damage to the structure of the aircraft. In the case of a crash, all parameters can significantly change requiring some serious actions on part of the passengers in the helicopter with time being the most critical factor.

9.1 Regulation of Helicopter Safety

In the UK, helicopter travel is regulated by the CAA (Civil Aviation Authority) Safety Regulation Group (SRG) and HSE's Offshore Safety Division (OSD) where the former regulates the legislation related to aviation and the latter deals with health and safety legislation.

CAA's Air Navigation Order 2009 (ANO) (SI 2009 No. 3015) requires duty holders on offshore installations to cooperate with helicopter operators in the provision of suitable helidecks. The Helideck Certification Agency (HCA) certifies helidecks to be safe for helicopter operations. Flight crew training, testing, licensing and working hours are regulated by the CAA. Legislation such as the Air Navigation Order is published by the CAA in documents referred to as Civil Aviation Publication or CAP.

A memorandum of understanding (MOU) between HSE and CAA determines the duties and regulations related to offshore travel. This MOU helps avoid duplication of effort in areas of overlap between the two organisations. Whilst HSE has a general responsibility for health and safety ashore and on-board offshore installations, their role is taken over by CAA for these matters on-board aircraft. For example, CAA standard for 'offshore helicopter landing areas' known as CAP 437 has to be complied with by the duty holders to provide safe helidecks. However, it is the HSE that will implement this requirement on offshore installations

Figure 9.2 Helicopter on a normally unmanned installation (NUI)

Figure 9.3 Helicopter seats and exits

where the duty holder will need to submit a safety case in align-ment with CAP 437. If HSE is satisfied with the safety case, only then the HCA will be able to issue the required certificate of compliance.

Generally, the duty to travel safely[2] is placed on all stakeholders involved in ensuring safety of offshore travel i.e.

1. helicopter operators;
2. flight crews;
3. installation operators;
4. offshore workers/passengers.

In order to ensure that helicopter operators comply with CAA requirements for passenger transport, they are required to obtain an Air Operator's Certificate (AOC). This certificate can only be issued if an operator produces procedures in the company's oper-ations manual. The operators are also required to provide a safety briefing at the heliport, certain PPE such as immersion/helicopter transit suits, EBS/inflatable lifejacket and ear protection.

In addition to the responsibility of enforcing Air Navigation (Dangerous Goods) Regulations 2002 (SI 2002 No. 2786), the CAA also has a responsibility to oversee the audits of training for offshore workers who:

1. prepare dangerous goods consignments for transport by air;
2. accept transport of cargo by air;
3. process passengers;
4. load baggage or cargo.

HSE will carry out incident/accident investigations[3] that occur within its area of responsibility. Through the MOU, HSE and CAA will cooperate with each other as well as other organisations such as DfT's Air Accident Investigation Branch (AAIB). All aircraft accidents will primarily be investigated by AAIB but HSE may also conduct parallel investigations for accidents within its areas of responsibility.

Duty holders on-board offshore installations have a responsi-bility to ensure:

1. safety on the helideck including the safety of personnel during helicopter landing, takeoff and fuelling operations;
2. suitability of the helideck at all times;
3. competent personnel are available to deal with any helicopter emergencies and ensure rescue of personnel if required;
4. adequate communication between various personnel during hel-icopter operations;
5. availability of a weather policy for helicopter passenger safety.

Due to recent helicopter accidents in the North Sea, the CAA, Norwegian CAA and European Aviation Safety Agency (EASA) conducted a comprehensive review[4] of offshore helicopter operations. The recommendations included a change in training requirements (see EBS section) and in addition:

- a ban on helicopter flights in severe weather conditions to avoid hindrance in rescue operations and to limit the possibility of helicopter capsize after ditching;
- improved standards for pilot training and checking;
- improved regulatory control for helicopter certification and airworthiness through European Aviation Safety Agency;
- standardisation of helicopter operating procedures throughout the sector.

9.2 Helicopter Systems and Equipment

Helicopter safety has always been given due attention since their use for commercial flights for the offshore industry in the early 1980s. Systems such as HUMS (Health and Usage Monitoring System), Advanced Anomaly Detection (AAD) for HUMS, Traffic Collision Avoidance System (TCAS), Helicopter Operations Monitoring Programme (HOMP) are some examples of the systems that have been utilised to improve the safety of helicopters. These systems are described below for readers to give them some background information to increase their confidence in travelling by these rotary wing aircraft.

9.2.1 Health and Usage Monitoring System (HUMS)

This system was introduced to the offshore oil and gas industry in the 1990s and subsequently expanded into the military and other operations. The system was designed to provide a means for early detection of problems that developed during a flight or after carrying out routine or reactive maintenance. The problems identified include progressive defects due to faulty installation, deterioration of bearings or gears, imbalances in equipment or other problems at an initial stage by utilising trends developing in parts of a helicopter structure. These problems are identified by using a large number of sensors (25 to 70) throughout the airframe and other components particularly all vibrating or spinning parts.

In older systems, this data can be downloaded and analysed by using specialised software after the flight whereas in modern systems, the on-board computer can perform the required analysis to inform the crew with an option to transmit this data via satellite to the ground based station. These sensors gather data and record it onto a memory card with the aid of a small on-board computer. In a typical system, the hardware includes a main computer or processing unit, an optical tracker, a display unit for the cockpit and a data transfer unit which typically houses the memory card to manually transfer the data.

In all cases, the information provides an early warning of any potential problems which can be rectified to avoid flight delays, downtime and therefore enhance safety in addition to improving profitability.

Figure 9.4 Helicopter on an offshore installation

Table 9.2 – Main functions[5] of HUMS	
Function	**Monitored Areas**
Transmission Vibration Monitoring	Engine to main gearbox shafts, gearbox shafts and gears, bearings, tail rotor drive shafts and bearings, oil fans
Usage Monitoring	Logbook data to include: • Engine starts, ground idle and running time • Rotor turning, operation and start stop time • Flight time and landing count
	Transmission Usage Monitoring (TUM) to record rotor speed, engine torque, main and tail rotor torque
	Structural Usage Monitoring (SUM) to include roll and pitch attitude, acceleration, altitude, vertical speed, air speed, roll and pitch rate
Rotor Track and Balance	Monitors vibration of the main and tail rotor

9.2.2 TCAS – Traffic Collision Avoidance System

Traffic Collision Avoidance System (TCAS)[6] also known as traffic alert and collision avoidance system or Airborne Collision Avoidance System (ACAS) is a system that assists in avoiding Mid

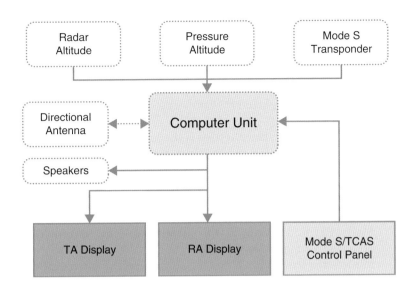

Figure 9.5 TCAS basic components

Air Collisions (MAC) between aircraft. The system monitors the airspace around the aircraft by detecting signals from other aircraft with TCAS enabled transponders.

Generally Air Traffic Control (ATC) provides separation of air traffic flowing in different directions through the allocation of airspace combined with the position supplied by pilots. However, due to mid air collisions, a need to establish on-board capability to warn of a risk of collision was seen in the 1950s. As a result, the predecessor systems of TCAS were developed. An example of such a system is BCAS (Beacon Collision Avoidance System) that used ATC beacons to establish any conflicting traffic's range and altitude. Whilst this system proved useful for some time, another accident in the 1970s led to further improvements in the system, which today is known as TCAS.

TCAS operates independently of the ground based systems such as ATC and provides information about other craft in the vicinity in real time. The signals from transponders are transmitted on a frequency of 1030 MHz and reflected on 1090 MHz frequency between the transponders of the two aircraft. In theory, one aircraft transmits signals, looking for another aircraft in the vicinity and if the transmitted signal interrogates another aircraft, the receiving transponder replies on 1090 MHz frequency. This cycle repeats several times in a second ensuring multiple transmissions and reception of signals.

The received signals allow the computer to establish the range, altitude (if transmitted in the reply) and bearing for the traffic. On the basis of this information, the Closest Point of Approach (CPA) is established with two types of alerts made available to the pilot:

1. Traffic Advisories (TAs) that help the pilot identify conflicting craft and prepare for evasive action if necessary;
2. Resolution Advisories (RAs) that recommends manoeuvres to the pilot to maintain or increase the current CPA. The operator/ pilot has the ability to select TA or TA/RA modes where the former uses mode A and the latter uses mode C (see below).

The 'Mode' for surveillance system is the format of pulses from an interrogating or a similar system. The reply format is referred to as a code. The modes are:

1. Mode A: provides air-to-air data exchange with identifying code only.
2. Mode C: in addition to the identifying code, information exchange also includes altitude.
3. Mode S: the letter S has been taken from Selective Interrogation, hence Mode S. This function avoids clogging the airspace when a large number of TCAS transponders interrogate each other. Consequently, mode S transponders transmit their identity at regular intervals. The identified targets in the area are recorded in the internal database. Only selected targets are then interrogated to provide air-to-air data exchange between the two aircraft. The data exchange information contains aircraft identity, altitude and target capability of the transponder and flight status (airborne/landed). Failure of this transponder will put TCAS into standby mode and warn the operator. Mode S is backwards compatible with Modes C and A, i.e. a Mode S equipped TCAS can always interrogate Mode C/A transponders.

TCAS is designed to operate with three variants i.e. TCAS I, TCAS II and TCAS III for different types of craft, collision avoidance requirements and obviously the cost. TCAS I is meant to be used by the smaller aircraft, is least costly and designed to allow the operators to visually acquire the nearby traffic. Its detection range is 4 nautical miles, detecting targets fitted with Mode C or S transponders and providing range, bearing and altitude. For mode A transponders, only range and bearing information can be obtained. TCAS I can only provide visual and audio RAs within 40 seconds of coming within a preset CPA. No guidance for evasive manoeuvres to avoid collisions is provided.

TCAS II performs all functions for TCAS I but in addition also provides RA to avoid CPA conflicts. Its range as well accuracy for range and bearing is greater than TCAS I. However, TCAS II RAs are restricted to the vertical plane. TCAS III on the other hand performs all functions for TCAS II and in addition has the ability to issue RAs for both the horizontal as well as the vertical plane. TCAS III has not been commercially successful due to the small number of benefits available in this progression; even though directional antenna was experimented in this version, it provided limited accuracy, hence it couldn't prove very successful. Further research is underway to develop this system and make it commercially available.

9.2.3 Glass Cockpit

Today's helicopters feature the most modern 'glass cockpits' fitted with electronic displays that utilise touch screen LCD displays for various controls that reduce clutter and provide information in a comparatively much smaller area. A more modern term for these systems is known as 'Integrated Cockpit Display System' or ICDS. On the contrary, the 'steam cockpit' i.e. the old style analogue gauges and dials has now almost become obsolete except in some older aircraft which are also being retrofitted with glass cockpits. However, key gauges such as altitude indicator or rotor rpm indicator are still kept to provide back-up systems.

The history of glass cockpits goes back to the 1970s when CRT (Cathode Ray Tube) displays were used which did not allow much on screen control as provided by today's technology. However, as air travel became more common, air traffic increased and the airspace became more congested, the need to make quick decisions also increased requirements for quick analysis of the available information. Additionally, the avionics industry also had to keep pace with advancements in other sectors such as the developments in the GNSS (Global Navigation Satellite System) positioning systems such as GPS (Global Positioning System), GLONASS (Globalnaya navigatsionnaya sputnikovaya sistema) and Galileo combined with modern radars etc.

The modern glass cockpit forms part of the EFIS (Electronic Flight Instrument Systems). A typical glass cockpit consists of the following segments:

- primary flight symbology;
- engine and system data;

Figure 9.6 Glass cockpit (left), steam or analogue cockpit (right)

- flight management data;
- digital maps;
- radar display;
- 3D Helicopter Terrain Awareness Warning System (TAWS);
- flight recorder.

The benefits of these systems include selection of the required information at the flight crew's choice to enable better situational awareness, reduction in their workload and automatic recording of all information in a database. Salient features of these systems are:

- customisable display supported by configurable software allowing the most relevant information to be displayed;
- low maintenance costs in comparison with analogue displays;
- compact design allowing a reduction in the cluttering of the cockpit and providing more space for other purposes;
- reduced weight due to use of smaller electronic components allowing more carriage capacity;
- redundancy of controls to provide back–up in case the electronic controls fail.

9.2.4 Minimum Equipment List (MEL)

Helicopters are required to comply with the regulatory requirements for carriage and fitting of minimum equipment[7] before they are given a licence to operate under the ANO. This list is referred to as Master Minimum Equipment List (MMEL) and is specific to a particular make/model of the helicopter recommended by the manufacturer.

The MMEL establishes the specific equipment on an aircraft that can be inoperative whilst it is allowed to function safely within defined conditions. These conditions, limitations and associated procedures will be described in the MMEL.

However, at times due to operational requirements, the operator may need to restrict some of the equipment which is listed in the Minimum Equipment List (MEL). The difference in MEL and MMEL may only be because some equipment has been put out of use due to operational requirements such as route, geographical location, availability of spare parts and maintenance capabilities. It is therefore important to note that the aircraft's approval from the likes of authorities such as CAA (UK), Transport Canada or FAA (USA) will be for MEL and NOT the MMEL.

MEL must also be approved by the likes of CAA. According to the ICAO (International Civil Aviation Organization), MEL is a document that:

- identifies the minimum equipment and conditions for an aircraft to maintain its certificate of airworthiness;

- defines operational and maintenance procedures to:
  - maintain the level of safety;
  - secure and deal with equipment that has been made inoperative.

When an MMEL is updated, the corresponding MEL must also be updated to ensure continued compliance or else the aircraft can be rendered unairworthy.

9.2.5 Emergency Equipment

The ICAO has established three classes of helicopters on the basis of their performance. These are:

1. Helicopter Performance Class 1 – A helicopter which due to critical power unit failure is capable of landing safely after a failed takeoff or continue with its flight to a safe landing area depending upon when the failure occurred.
2. Helicopter Performance Class 2 – A helicopter which due to critical power unit failure is capable of continuing its flight safely to a safe landing area except when the failure occurred during takeoff or late in the landing manoeuvre in which case it may need a forced landing.
3. Helicopter Performance Class 3 – A helicopter which due to critical power unit failure at any time during the flight will require a forced landing.

On the basis of the above classes of helicopters, different types of emergency equipment[8] may be required by helicopters and all personnel travelling in them as described below.

1. Helicopters flying over water must be fitted with permanent or rapidly deployable means of floatation for safe ditching when engaged in offshore operations beyond a distance from land that equals more than 10 minute cruising at normal speed for performance class 1 or 2 helicopters and beyond safe forced landing distance from land for class 3 helicopters.
2. Performance class 1 or 2 helicopters flying over water should be equipped with:
   a. one lifejacket or equivalent floatation device worn by each person on-board when engaged in offshore operations unless the person is wearing an integrated survival suit fitted with built-in lifejacket;
   b. life rafts (helirafts) of capacity suitable for number of persons being carried stowed in such a way that they can be used immediately in emergency. If two liferafts are fitted, then each must provide a capacity equivalent to the total number of persons carried by helicopter;

c. distress signals as given below:
   i. SOS (. . . _ _ _ . . .) made by any signalling method;
   ii. radio telephony or data link signal consisting of word MAYDAY;
   iii. rockets or shells throwing red light fire one at a time;
   iv. a parachute flare showing a red light;
   v. setting of transponder to Mode A Code 7700.
3. For offshore operations, a survival suit must be worn by all occupants when the sea temperature is less than 10°C or when estimated rescue time is likely to exceed the calculated survival time.
4. The pilot must ensure that each person in the helicopter is familiar with:
   a. seat belts;
   b. emergency exits;
   c. lifejackets (for non-offshore flights as these will be stowed and not worn by personnel);
   d. oxygen dispensing equipment;
   e. other emergency equipment.
5. All helicopters shall be fitted with at least one Emergency Location Transmitter (ELT).
6. A helicopter flying at altitudes of atmospheric pressure less than 700 hPa shall be equipped with oxygen storage and dispensing apparatus.
7. All helicopters operating in icing conditions must be equipped with anti-icing and/or de-icing devices.
8. Helicopters carrying passengers must be fitted with weather detection equipment.

9.3 Helicopter Crew Training, Qualifications and Checks

It is important to make a distinction between various roles that may be allocated to the crew of helicopters operating for offshore transport:

• Aircraft Commander, Pilot in Command (PIC) or simply pilot is the person in charge of safe operation of the aircraft and safety of all personnel on-board.
• Co-pilot is a qualified pilot but may have less experience than the PIC and assists the pilot in ensuring helicopter safety.

In the aviation industry, the Airline Transport Pilot Licence (ATPL) is required to be the commander of a craft of over 5700 kg or with more than nine passenger seats. The classifications for helicopter pilot are ATPL(H) or Commercial Pilot Licence [CPL(H)].

Table 9.3 – Qualifications and experience required for an ATPL(H)[9]	
	Helicopter PIC qualifications
Total flying hours	1000
Multi-pilot operations	350
As PIC or	250
As PIC/PIC under supervision or	100/250
As PIC under supervision (will be restricted to multi-pilot operations until 100 hours PIC completed)	250
Cross-country flight time with 100 as PIC	200
Instrument time	30
PIC/Co-pilot of night flight	100

Experience not less than (label for left side of table)

Table 9.4 – ATLP(H) theory modules		
Module 1	**Module 2**	**Module 3**
• Aircraft General Knowledge – Instrumentation • General Navigation • Meteorology • Human Performance	• Radio Navigation • Aircraft General Knowledge – Airframes, Systems and Power Plants • Air Law • Flight Planning and Flight Monitoring	• Operational Procedures • Principles of Flight • Performance • Mass and Balance • Communications – VFR (Visual Flight Rules) and IFR (Instrument Flight Rules) Communications

In order to obtain an ATPL(H) licence, the requirements are given in Table 9.3. In addition, the candidates need to undertake a module on Maths and Physics in some cases. The classroom theory divided into three modules given in Table 9.4 requires 550 full time equivalent hours. A pass in the CAA examination at the end of each module is required to progress to the next module. All modules must be passed within 18 months of undertaking the first module in a total of six sittings with an allowance of five attempts for any one module.

In order to obtain a CPL(H) licence, the following requirements must be complied with:

- at least 18 years of age;
- hold an ICAO compliant Private Pilot Licence PPL(H);
- complete an approved course;
- pass all required examinations;
- obtain the flying hours required depending upon the route followed. For example, if a modular route is taken, then a total of

155 flying hours are required within which 35 hours is spent during training for CPL(H).

The total duration of the CPL(H) course includes 500 hours of classroom teaching followed by written examination in the following modules:

- Air law and ATC procedures
- Aircraft general knowledge
- Flight planning
- General navigation
- Human performance and limitations
- IFR (Instrument Flight Rules) communications
- VFR (Visual Flight Rules) communications
- Instrumentation
- Mass and balance
- Meteorology
- Operational procedures
- Performance
- Principles of flight
- Radio navigation.

A comparison of the required experience and qualifications for pilot and co-pilot operating in the offshore industry are given in Table 9.5.

For each course discussed above, a considerable amount of time is spent on developing the candidate's skills and knowledge for soft

Table 9.5 – Comparison of qualifications and experience of PIC and Co-pilot for offshore operations[10]		Helicopter PIC qualifications	Co-pilot qualifications
Licences		ATPL(H)	CPL(H)
Type rating on contract aircraft		Current	Current
Instrument rating on contract aircraft		Current	Current
Experience not less than	Total hours	2000	500
	Total hours in command	1000	250
	Total hours in command – multi eng	500	
	Total hours in similar aircraft complexity	500	100
	Total hours command on contract type	100	50

skills such as communication, behaviour, organisation, airmanship and crew cooperation. They are given the highest possible practice on flight simulators dealing with various simulated emergency scenarios, using controls and instruments prior to putting these skills into practice under supervision, using their flying hours. Once trained on a specific type of helicopter, they may require further training referred to as 'type rating training' on specific instrumentation and/or different types of helicopter on which they may be deployed.

Once qualified, pilots are required to undertake recurrent training on an annual basis to ensure they maintain their competencies by using classroom based, simulator, emergency, survival and other relevant training. In addition, they are also required to go through flight checks covering the exercises as applicable to the type of helicopters they are likely to fly. A separate check is required for day and night for each type of helicopter.

9.3.1 Air Operator's Certificate (AOC)

In the UK, in order to fly commercial aircraft as public transport, the operator or pilot must be issued with an Air Operator's Certificate (AOC). Passengers can check that the operators hold AOC because this certificate signifies that the operator holds the required competency to operate the craft safely. The ICAO requires a certified true copy of the AOC to be carried on-board the aircraft.

9.3.2 AOC Suspension and Revocation

AOC may be suspended by the CAA upon request of the holder or if the holder is unable to meet the requirements. The latter type of suspension may be temporary until the holder meets the requirements again within a period of 3 months. For durations longer than 3 months, the CAA will start considering a revocation process until the holder provides a planned action to return to operations. However, after a period of 6 months from suspension, the AOC will be considered to be revoked. Subsequently, a fresh application for AOC will be required to gain the same status.

9.3.3 Crew Health

1. Crew members of an aircraft are forbidden from being under the influence of alcohol and drugs. The legislation requires them to abstain from consumption of alcohol for at least 8 hours prior to their duty.
2. Use of medication such as antihistamine has been known to contribute to accidents. Similarly, melatonin which is available as a food supplement in many countries is known to impact

on circadian rhythm. Primarily it is used to overcome problems such as jet lag but research carried out by the CAA indicated that it caused nausea, depression and other problems. As a consequence, it is recommended not to consume melatonin in the 12 hours prior to flying.

3. Airline transport pilots are required to have a Class 1 medical certificate prior to being issued with a pilot's licence or for undertaking training. For a multi-crew pilot licence, individuals must be at least 18 years old whereas for an ATPL, they must be at least 21 years old. The medical examination undertaken prior to the certificate being issued can take up to 4 hours and includes examination of the following:

   a. medical history to check about any history of previous illness that may affect an individual's ability to fly commercial passenger transport aircraft;
   b. eye test to check the eyesight;
   c. physical examination of the body to include heart, blood pressure, limbs and nervous system to ensure correct functionality;
   d. a basic hearing test to evaluate hearing;
   e. lung function test to check the ability of lungs to exhale air, for example due to asthma;
   f. blood test to check blood's ability to carry oxygen i.e. lack of haemoglobin causing anaemia;
   g. urine test for diabetes or blood in urine.

9.4. Helicopter Maintenance

Due to the nature of operations, helicopters require a high degree of maintenance to ensure the safety of operations and airworthiness of the craft. From 2000 to 2005, the European Helicopter Safety Team (EHEST)[11] found that 14 per cent of accidents were linked to maintenance related issues. In some cases, lack of follow-up for maintenance recommendations led to the accidents. EHEST also identified that if the Intervention Recommendations (IR) were implemented, a number of incidents could have been prevented or their severity reduced. As a result of their studies, EHEST divided IR into the following categories:

- adherence to recommendations for continued airworthiness;
- compliance for quality control and worker qualifications;
- company support for a management system, quality assurance and engineering;
- competence assurance for workers and technicians;
- availability and utilisation of OEM (Original Equipment Manufacturer) information and use of approved parts;
- oversight of all aspect of maintenance by the CAA.

Helicopter maintenance can be split into various areas such as unscheduled/corrective, progressive/preventative or base maintenance. When a helicopter needs to be transferred from one place to another, purely for maintenance purposes and where this transfer requires a flight, no commercial passengers or cargo should be carried in the craft. EASA regulation EC 2042/2003 part 145 deals with approval of organisations to undertake helicopter and other aircraft over 5700 kg, multi-engine helicopters or commercial aircraft's airworthiness and maintenance. Various types of helicopter maintenance are discussed in this section.

9.4.1 Unscheduled or Corrective Maintenance

This type of maintenance is required for repairs of damaged, faulty or failed components or parts of a helicopter to bring it back into full service. Health and Usage Monitoring System (HUMS) can be of significant assistance in establishing the main and associated faults to ensure a complete rectification of the problem. Modifications, installation of new systems or components form part of corrective maintenance.

9.4.2 Progressive or Preventative Maintenance

This type of maintenance is required to prevent or reduce the probability of failure after a predetermined period of time such as that recommended by the manufacturer. Preventative maintenance is divided into four types:

1. Servicing: Regular servicing required after a set period of use such as total flying hours or the usage life of individual components.
2. Scheduled maintenance carried out at set regular intervals such as daily, weekly, monthly or annual periods.
3. Condition based maintenance: Condition of certain components is monitored on a continuous basis. When it reaches a predetermined physical condition or its output reaches a predetermined level, then maintenance of the required component is carried out. Methods such as non-destructive testing by using X-rays or other techniques may be used to establish the condition of a particular component.
4. Out of phase maintenance: The maintenance of many components may not synchronise with the general maintenance programme. Any such maintenance will come under out of phase maintenance.

Whilst most of the maintenance will be carried out by licensed engineers or technicians, pilots who hold maintenance qualifications (see next section) are allowed to carry out preventative maintenance provided their rest hours are not compromised. Generally, the pilots can carry out the daily pre-flight and turbine engine power assurance check. Qualified engineers will carry out timed checks such as 50/100 hour inspections or annual inspections etc.

9.4.3 Base Maintenance

Line maintenance tasks include pre-flight maintenance to ensure an aircraft's airworthiness. Typical examples of line maintenance tasks include trouble shooting, defect rectification, component replacement which may include replacement of engine or propellers, scheduled maintenance or checks to detect unsatisfactory conditions that do not require in-depth inspection for structures, systems or power plant visible through quick opening access panels. Any tasks that can't be performed under line maintenance must be performed in base maintenance.

Base maintenance requires the following specialised facilities along with appropriately qualified staff for all areas of work:

- hangar(s) of suitable size to accommodate the required type of aircraft;
- specially equipped workshops where aircraft can be protected from weather elements during maintenance;
- office accommodation for planning and managing the work being undertaken as well as maintaining records;
- storage facilities for equipment, components and tasks;
- quality assurance/control and management systems;
- the staff required for base maintenance to include:
  - base maintenance manager;
  - line maintenance manager;
  - workshop manager;
  - quality manager;
  - qualified technicians, mechanics and engineers to carry out required maintenance.

Examples of maintenance tasks that can be carried out include:

- airframe inspections and checks;
- modifications and refurbishment of airframes or other components;
- avionics inspection, maintenance or upgrade;
- equipment calibration, testing and recertification;
- Non-Destructive Testing (NDT) to check integrity of components;
- machining;
- welding;
- painting.

9.4.4 Types of Maintenance Engineer

Also known as aircraft engineers or technicians, these personnel employed for the maintenance of helicopters require appropriate qualifications that enable them to carry out maintenance on the type of craft for which they are qualified. Where newly qualified personnel are employed, the employer is required to ensure adequate level of supervision at all times. The same rule applies for trainees who are also required to work with a formal training plan.

The European Aviation Safety Agency (EASA) regulates all aviation activities including training in the EU with the UK's CAA acting on EASA's behalf to monitor all activities locally. EASA's Implementing Rule (IR) part 66 is a common standard utilised in Europe and many other countries for licensing aircraft maintenance personnel.

The aircraft maintenance personnel are qualified in the same way as pilots or Air Traffic Controllers (ATCs). Fixed or rotary wing aircraft engines are of two types i.e. turbine engines or piston engines. Turbine engines are also known as jet or combustion engines and utilise turbines but they may also be propeller driven. The piston engine, also known as a reciprocating engine, on the other hand utilises pistons within a cylinder that operate on combustion of fuel to move the propeller shaft, hence the aircraft fitted with piston driven engines always use propellers. Various qualifications for helicopter maintenance engineer are described below on the basis of engine type utilised.

Two types of licences are:

- Category A: This category applies to the line maintenance certifying mechanic for airframes and engines and is subdivided into the following categories:
  - Cat A1 – Turbine Engine Fixed Wings
  - Cat A2 – Piston Engine Fixed Wings
  - Cat A3 – Turbine Engine Rotary Wings
  - Cat A4 – Piston Engine Rotary Wings
  - Cat A5 – Airships.

Figure 9.7 Helicopter being towed to hangar for maintenance

- Category B: Applicable to line maintenance, base maintenance in inspection or supervisory roles. This category is subdivided into B1 and B2 for mechanical and avionics disciplines respectively. These categories are further subdivided into the following specialities:
  - Cat B1.1 – Turbine Engine Fixed Wings
  - Cat B1.2 – Piston Engine Fixed Wings
  - Cat B1.3 – Turbine Engine Rotary Wings
  - Cat B1.4 – Piston Engine Rotary Wings
  - Cat B1.5 – Airships
  - Cat B2 – Line Maintenance Certifying Technician.

For achieving the required qualification for both Cat A and Cat B maintenance engineer or technician, personnel holding GCSEs in maths and science can undertake modular courses leading to the above qualifications. At the end of each module, the candidate undertakes an examination valid for a period of 10 years (5 years for Cat A qualifications). Once a pass is achieved for all required modules along with the required maintenance experience, the candidate will gain the qualification according to the modules studied. Generally on the job experience level varies from 2–5 years for Cat B qualifications and 3 years for Cat A qualifications. Upon achieving the required qualification, a person becomes a Licensed Aircraft Engineer (LAE) for work on a specific type of aircraft in the relevant area such as airframe, engine or avionics. Examples of qualifications are:

1. A Line or Ramp Maintenance Certifying Mechanic: tasked to carry out minor maintenance tasks or repairs that do not require disassembly. For example, replacement of parts during the service life of an aircraft. Upon completion of the task, the mechanic certifies its completion.
2. Base Maintenance Certifying Technician: tasked to work on an aircraft that has been taken out of service for routine maintenance, major overhaul or refit. Upon completion of the task, the technician certifies its completion.
3. Unlicensed Maintenance Engineer – cannot certify completion of the task to release the aircraft for service. However, they can assist the licensed engineers in undertaking the task. An engineer can be unlicensed who has achieved the basic qualification to become an aircraft engineer but requires the mandatory experience to gain a licence. This may apply to trainees or apprentices who are working towards their qualifications for obtaining LAE.

In addition to the above, helicopter operators may also have

- Maintenance or Chief Engineer: Their job includes resource managing helicopter maintenance jobs, planning and ensuring availability of required personnel, equipment and spare parts. This person will be an experienced aircraft engineer. They will also liaise with organisations such as CAA for review and audit of current practices and procedures.

- Airworthiness Manager: This role will involve development of the maintenance programme as well as following guidance from CAA such as airworthiness directives or service bulletins from the manufacturer to ensure the aircraft's airworthiness is ensured at all times. In doing so, they will ensure all records of maintenance are kept up to date and made available during audits and inspections.

9.5 Role of Passengers in Helicopter Safety

Whilst the helicopter operator and crew are required to ensure the airworthiness of the craft, passengers play an important role in ensuring their own as well as the helicopter and its crew's safety. Some considerations for passengers in this respect are as follows:

- Passengers must not attempt to carry banned items such as sharp objects, liquids, etc. on-board.
- Passengers must follow crew instructions at all times particularly paying attention to the pre-departure briefings. They must

Figure 9.8 A passenger carrying his bags to board the helicopter

familiarise themselves with the liferafts, procedure to jettison emergency exits or pop out windows and reference points for escape from helicopter in case of ditching at sea.

- Passengers must don fully the safety harnesses or belts, life-jackets, ear protection and survival suits as required under the procedures for offshore flights. None of these items should be loosened or removed during the flight.
- Restrictions on smoking (including e-cigarettes), chewing tobacco and use of alcohol must be strictly followed at all times.
- Carriage of prohibited items such as sharp objects, liquid fuels, matches or lights must be complied with at all times.
- Pre-flight information will advise the passengers about restrictions on hand carried baggage. This must be strictly complied with to avoid overloading the craft and/or causing other safety risks.
- Electronic devices including mobile phones must be switched off during takeoff, landing or flights. Any doubts such as use of a device in 'flight safe' mode must be checked with the crew prior to using it.
- The pilot has the authority to refuse boarding to any passenger whose presence could present a hazard to the safety of other passengers, crew or the aircraft, hence all passengers must avoid taking any chances by lack of compliance with crew instructions.
- When at the heliport or helipad, passengers must restrict their movements to the allocated zones only. They must not walk into, for example, the fuelling zones.
- Passengers must remain in their seats with their seat belts/harnesses fastened for the whole duration of the flight.
- Any passengers travelling offshore who are under the influence of alcohol may be refused boarding.
- Passengers must not disembark until instructed by the pilot, helideck landing officer (HLO), etc.
- Passengers must approach the helicopter from the sides and never approach the helicopter from the rear. The furthest they should go is to the baggage compartment to collect their baggage when instructed by flight crew or the HLO. Helicopters should not be approached by passengers during the start-up or shutdown.
- Hand carried objects such as sunglasses, caps or other objects must be securely held to prevent being blown off by the wash caused by the main rotor.
- In an emergency such as helicopter landing/ditching over water:
  - passengers must not evacuate the craft until the rotor has stopped;
  - they must not inflate their lifejacket until outside the craft.
- Be aware of the type of seats used such as stroking or crash attenuating seats and the risk of trapping hands or fingers under the seat when it decompresses due to sudden movement.

9.6 HUET Training Areas Covered

Helicopter Underwater Escape Training (HUET) has been undertaken by personnel flying on helicopters since the 1970s. Issues that need to be addressed during HUET training are based on the difficulties faced by survivors following an accident involving a helicopter landing on water, capsizing or submersion. A common factor amongst all helicopter ditching is the fact that once it is on the surface of water, it is unlikely that it will remain upright due to being top heavy. The water ingress will continue to increase load within it and eventually it may sink. Consequently, the passengers may be trying to escape from a helicopter which is continuously in motion.

Helicopters are designed to remain afloat in an upright position for a time period that will allow passengers sufficient time to escape in a Beaufort wind force scale 4 i.e. wind speed 11–16 knots, wave height 1–2 m. However, in the North Sea in particular and in other areas in general, the sea state exceeds this scale quite regularly. As a consequence, research continues into methods to improve the stability of helicopters to provide additional floatation and thus allow more time to escape in an emergency. Until a full, better alternative is found, the current system must be followed comprehensively to ensure all known risks are mitigated. The last resort is to know the issues and keep them in mind if the need to escape in an emergency ever arises. These and other factors that make an egress quite a challenging task are summarised below:

1. Difficulty faced during egress at the time when water is rushing into the helicopter.
2. Difficulty faced in reaching and opening the exits.
3. Ability of the helicopter to remain afloat dictating the time during which passengers have to escape.
4. Disorientation, especially if the helicopter capsizes due to failure of inflation equipment to support helicopter in an upright position.
5. Panic resulting in survivors' inability to take required actions to abandon helicopter.
6. Time of the day or night, actions that could be taken easily may be difficult to take at night.
7. Temperature difference between air and the water in which a helicopter ditches. Large difference in temperature could lead to cold water shock which may increase the shock caused by the accident. Consequently the endurance provided by EBS (Emergency Breathing System) will be reduced due to rapid breathing.
8. Cold water shock could lead to drowning or hypothermia if the required PPE is not supplied or not used appropriately.

Figure 9.9 Dry evacuation into a heliraft (Source: ©Petrofac Training)

9. Possibility of the exit being blocked by another person and or objects that may float free which may also give rise to entanglement issues such as from lanyards, ropes or straps, etc.
10. Injuries caused during ditching or subsequent damage to the helicopter.
11. Impact of weather elements such as wave or swell on escaping from the helicopter.
12. Inability to operate equipment as required such as releasing seat belts, using EBS or operating release mechanisms for exit windows.
13. Ability to use survival equipment.

Whilst no two ditchings will be the same, it is the above factors that are considered during the HUET training which make a significant contribution towards the ability of helicopter passengers to survive. It is important to note that various terminologies are used in the industry with reference to HUET training such as:

- HUET – Helicopter Underwater **Escape** Trainer
- HUET – Helicopter Underwater **Egress** Trainer
- HUET – Helicopter Underwater Escape/Egress **Training**
- HUEBA – Helicopter Underwater Emergency Breathing Apparatus
- HEBE – Helicopter Emergency Breathing Equipment
- HEED – Helicopter Emergency Egress Device.

In further description of the topic, this book uses HUET training as a standard term which encapsulates the terms described above. Various components used in the HUET training are described in the next section.

9.7 Emergency Breathing Systems (EBS)

An Emergency Breathing System (EBS) is a 'last resort' device to breath underwater in cases where the wearer needs to escape from a sunken helicopter. This system is also referred to as Emergency Underwater Breathing System (EUBS). However, this book will use the term EBS in the context of this system and its variants.

EBS have been used in the helicopter transport industry for many years due to their ability to provide breathable air for short periods that allow safe egress from the helicopter. If there are no significant factors preventing the escape, it is estimated that an EBS provides air for breathing for about 45–60 seconds to exit from a sunken helicopter. This duration can easily be reduced to around 10–15 seconds due to impact of shock and therefore panic, cold shock in waters of less than 10°C temperature and therefore rapid breathing. Generally, personnel can hold their breath for such durations but in the unlikely event where the exit is blocked or cannot be opened, holding their breath for longer durations will be impossible. Consequently, the chances of inhaling water will increase tremendously.

Helicopter operators for the offshore oil and gas industry use three types of EBS:

Figure 9.10 Viking helicopter transportation suit 4003 (Source: www.viking-life.com)

- Re-breather Systems – Commonly used in the North Sea sector, these system use a bag, commonly referred to as 'counterlung' into which the users exhale their own breath and then use it when required. The re-breather system is fitted with a mouthpiece which can be put in the mouth, in a position to start using the exhaled air by releasing an activation mechanism. However, this is not required until the user actually needs to use the air from the counterlung at the expiry of their breath-hold capability. Some re-breathers are fitted with automatic water activation mechanism. The limitations of this system are that it fills with water if submerged after activation rendering it useless and their effectiveness in only for a few metres (approximately 5 metres) after which the hydrostatic pressure makes breathing from the counterlung almost impossible. Because of these limitations, re-breathers are considered most suitable for controlled or semi-controlled ditching.
- Category 'A' EBS[12] – In this system, the air is supplied through P-STASS (Passenger Short Term Air Supply System) built into the lifejacket. The system will be approved by the UK's CAA and the European Aviation Safety Agency (EASA) and will meet the CAA's requirements below:
  a. capability to be used in air and under water;
  b. capability to be operated with one hand;
  c. simple to operate;
  d. only one action required to operate on submersion;

e. means provided to prevent water entry through nose;
f. mouth piece deployed within 10 seconds with full deployment of the system within 12 seconds;
g. air supply sufficient to breathe for 60 seconds at water temperature of 12°C.

- Compressed Air Systems – These can be compared with a Self Contained Underwater Breathing Apparatus (SCUBA) except that they are very small in size. Their components include a mouthpiece, demand valve, compressed cylinder and regulator. Breathing from compressed air can start 'on demand' as soon as the user activates the trigger. This design feature makes them superior to the re-breathers as they can be used at greater depths as well as when a helicopter sinks without any prior warning. In 2014, the UK introduced the use of Compressed Air Emergency Breathing System (CA-EBS) Initial Deployment Training for all offshore workers. This change resulted from the CAA report CAP 1145 recommending all personnel not seated next to an emergency exist to carry a CA-EBS from January 2015. These changes have been introduced to enhance safety of personnel travelling on offshore flights. Further details of this training are discussed in Chapter 11. CAA's study also included 'Development of a Technical Standard for Emergency Breathing Systems', CAA publication CAP 1034 published in May 2013 which defines standards for EBS design.

- Hybrid Systems – These consist of the combination of a re-breather and compressed air systems i.e. these systems are fitted with a re-breather counterlung and a compressed air cylinder. The cylinder capacity (usually 3.5 litres) is usually less than the compressed air EBS and releases air into the counterlung from where the user can breathe. The system is considered more robust than the other systems since if the re-breather system fails due to water ingress into the counterlung, then the compressed air can be used to breathe after purging water out of the bag. In addition, this type of EBS can be used in both ditching and helicopter sinking scenarios. It is worth noting that both the compressed air and hybrid EBS system carry an inherent risk of causing barotrauma – injuries caused by difference in pressure. Generally for EBS use, this is referred to as pulmonary barotrauma i.e. over-expansion of the lungs due to pressure of compressed air. The barotraumic impact can be even more severe if the user has to deploy an EBS at larger depths. The compressed air systems may also cause air or gas embolism i.e. trapping air in the blood stream and moving it to the heart or lungs. If a large air bubble is trapped in the heart, death may occur due to a heart attack as a result of lack of blood flow to the lungs or other vital body parts.

Regardless of which type of EBS is used, it needs to be located in a position where the user can rapidly deploy it. Various manufacturers have come up with slightly different ideas to locate EBS whilst being used with immersion suits. Examples of various types of EBS used with lifejacket or survival suit are given below.

9.7.1 LAPP Jacket

Until 31 December 2014, the UK offshore industry used Shark Group's (Survival One/Survitec Group) combined Air Pocket into the lifejacket, hence the name Lifejacket Air Pocket Plus (LAPP). This lifejacket kept the EBS at the chest of the wearer, reduced the number of items that had to be worn and therefore reduced the user effort in donning it as it could be worn with the survival suit. Approved by the CAA, a LAPP jacket provided 275 newtons buoyancy. It was also fitted with accessories such as whistle, buddy line, splash guard. The EBS was designed to automatically activate upon immersion into water. However, the LAPP jacket has now been superseded by the Mk50 Sentinel Lifejacket.

9.7.2 Survitec Group's Mk50 Sentinel Lifejacket

This is the most recently developed lifejacket for use by helicopter passengers travelling to offshore installations. The modified design of the lifejacket is based on CAA's report CAP 1145 published in response to the helicopter accident in the North Sea in August 2013. These changes became effective from January 2015 and require all passengers to use Category 'A' EBS as described earlier in this chapter. Salient features of this lifejacket include:

- twin chamber 275 newton inflatable lifejacket;
- spray hood;
- lifting loop;
- buddy line;

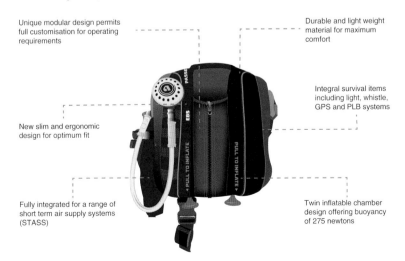

Figure 9.11 Survitec's new MK50 sentinel life preserver offshore helicopter passenger lifejacket (Source: http://apps2.survitecgroup.com)

Figure 9.12 Hansen helicopter transport suit (Source: www.hansenprotection.no)

Figure 9.13 Viking PS5040 aviation suit (Source: www.viking-life.com)

- compatible with helicopter transit suit;
- can be integrated with PLBs.

9.7.3 Hansen Transport/Immersion Suit – Sea Air

Manufactured by Hansen Protection AS, this suit is widely used in the Norwegian sector of the oil and gas industry. It is designed to be used for helicopter travel offshore as well as an immersion suit for abandoning an offshore installation. It is made from a waterproof and flame retardant outer layer and an insulated inner layer that can maintain body temperature to a comfortable level. The suit is fitted with a hood, spray hood, gloves, light, buddy line and lifting strap. The main feature of these suits is the integral PLB (Personal Locator Beacon) and re-breather type EBS.

9.7.4 Viking Transport/Immersion Suit – PS5040

Viking's helicopter transport suit is also a combined transportation and immersion suit approved by ETSO (European Technical Standard Order) and under SOLAS. It is fitted with a built-in lifejacket, integrated spray hood, neoprene boots and gloves. PLBs and EBS systems can be built into this suit.

9.7.5 Biardo Lifejacket Re-breather System and Immersion Suit

Biardo Survival Suits BV is a Netherlands based company specialising in helicopter transportation suits, lifejackets, re-breather systems and other safety products. The PTS003 helicopter transportation suit is European Aviation Safety Agency (EASA) and ETSO approved. The suit uses fire retardant materials on hood, gloves, and wrist and neck seals.

Biardo's LRS001 lifejacket re-breather system uses an EBS built into a 275 kN inflatable lifejacket. The re-breather lung can be filled with 4 litres of compressed air from the gas bottle.

9.7.6 Short Term Air Supply System (STASS)

Short Time Air Supply Systems (STASS) and Passenger Short Term Air Supply Systems (P-STASS), are compressed air systems consisting of a compressed air cylinder fitted with a mouthpiece either directly to the cylinder or through a hose. A reducer and a demand valve is fitted to control the pressure of air coming out of the cylinder. A nose clip is also provided to block breathing through the nose when STASS is in use. Primarily used by the crew of the helicopters, passengers can also use it if suitably trained in its use. The gas in the cylinder will generally be sufficient for about 2 minutes (0.5 litres) at a depth of about 5 metres but this time can vary depending upon usage which may change due to shock and therefore a change in breathing rate. The difference between STASS and P-STASS is that the latter does not require purging of the system by use of an occlusion system to avoid accumulation of water in the mouth piece.

Carried in a specially designed bag, it is positioned to the front of the body if supplied as personal equipment. Otherwise, these may be placed in strategic locations within the helicopters. When it needs to be deployed, users must insert the mouthpiece prior to the water level reaching their mouth or they are unable to hold their breath. STASS can also be deployed when underwater and does not require purging before inserting into the mouth. The order in which it should be deployed includes fitting the nose clip, placing the mouth piece in the mouth and gripping it firmly between the teeth forming a seal with the lips. Once required to breathe, just inhale/exhale normally from the mouth.

It is recommended that STASS/P-STASS should never be used without prior training. This is to avoid injuries resulting from barotrauma and/or air/gas embolism.

Figure 9.14 Escape from HUET using STASS (Source: Andark Diving)

9.8 Helicopter Underwater Escape Training Using EBS

Helicopter Underwater Escape Training has played a vital role in the survival of many helicopter casualties. This has only been possible through carefully planned exercises that take the attendees through various scenarios as required by OPITO standards. A brief description of this training is given here to give readers an understanding of the benefits if they were to face evacuation from a helicopter

Figure 9.15 Delegates practising with EBS in shallow water (Source: Warsash Maritime Academy)

in real life and also to know what to expect when they undertake training at a shore based training establishment. In order to ensure continued safety, trainees must draw the attention of trainers or other safety staff when they feel uncomfortable.

9.8.1 OPITO HUET Training

Various exercises during HUET training are described here to explain the requirements for HUET training. Readers must note that during HUET training, competent divers are available in the water along with other staff inside the HUET and supervisors on the side of the pool. The exercises are designed in such a way that even non-swimmers will be able to successfully and safely undertake them.

1. Preparation for ditching: In this exercise, candidates are trained to prepare for ditching on dry land. They are given a briefing on the procedure to follow when the helicopter has to land in an emergency. At the beginning of the exercise, all attendees are fully attired in their survival suits with the following four key steps:

   a. Belt: Four point harnesses fully buckled in and comfortably tight. Candidates will be given a full brief on pre-ditch actions including checking the seat belt, zipping up the survival suit (hood up zip up), identifying the nearest exist and then adopting the brace position.

   b. Brace: Adopt a brace position upon instructions from pilot. Depending upon location of the passenger i.e. in the window or side seat, a slightly different position may need to be adopted. Bracing for impact reduces the flailing when the helicopter falls under gravity and the secondary impact

Figure 9.16 Escape from HUET using STASS (Source: Andark Diving)

when it hits the land or water. Other general precautions include keeping the lower torso firmly against the seat, feet flat on the deck; shoes should be left on except high heeled shoes. Forward facing seats with shoulder harness (4 point harness), head should be tucked down as low as possible, one hand positioned on the buckle to release harness and the other hand positioned on the emergency exit. On helicopters fitted only with lap belts, the upper body should be bent forwards resting the head and chest against the thighs with hands holding the ankles. Consideration should be given to the direction of the exit and the release of the seat belt buckle. However, since all helicopters used in the offshore industry are fitted with 4 point harnesses, the use of this method is only for information to distinguish the procedures followed by passengers of commercial fixed wing aircraft. When stroking or crash attenuating seats are fitted, the brace position will be slightly different in that:

i. both feet should be flat on the floor of helicopter ensuring heels are not under own seat and toes not under the front seat;

ii. some distance kept between the feet to keep the legs separate to absorb downward jerk due to activation of the stroking seat;

iii. cross hands over the chest grasping the harness but ensuring thumbs are not under the harness;

iv. head should lean slightly forwards to avoid impact on the spine.

c. Buckle: On a four point harness, one hand should be kept on the buckle to release the harness immediately after the impact.

d. Exit: Passengers sitting next to an exit must keep one hand

Figure 9.17 Trainee seated in aircraft seat with lap belt only (Source: Ceronav, Romania)

Figure 9.18 Trainees preparing to exit from a partially capsized trainer (Source: Andark Diving)

on the exit whilst the others can keep their hand towards the exit to avoid being disorientated if the helicopter was to turn 180° after ditching.

2. Dry Evacuation: This drill simulates situations where the helicopter crew has some control over the helicopter landing and they can manage to prepare and use inflation equipment as soon as the helicopter lands on water or land. In this case, the main door(s) is used to exit into the heliraft. Generally, the heliraft is inflated in this drill where candidates exit directly into it. Here the instructor gives them an overview of the equipment found in the helirafts and subsequent actions that need to be taken.

3. EBS Acquaintance: On completion of the dry evacuation exercise, candidates undertake EBS acquaintance drill to familiarise themselves with the functions of the equipment, including its limitations. Compressed air is not used in the EBSs used for training (see Chapter 11 – CA-EBS). Candidates are advised to exhale their breath into the EBS counterlungs after taking a deep breath. Subsequently, they submerge their heads under water and breathe from the air in the counterlung whilst swimming to the side of the swimming pool. Important factors to understand during this exercise are the possibility of water ingress into the lungs if the mouthpiece is dropped in water after activation and an appreciation of the pressure of water on the counterlung when breath is being used to fill it.

4. Partial HUET capsize: Candidates are given an opportunity to familiarise themselves with the effects of helicopter capsize through various stages to build up their confidence and also to avoid undue stress in the training environment. The initial three stages are:

a. Without EBS without windows. In this drill, candidates are expected to exit a partially capsized helicopter through nominated exits which are not fitted with windows.

Figure 9.19 Trainee exiting from a fully capsized HUET (Source: Heli Subsea)

Figure 9.20 Trainees gathered in a heliraft after exiting from HUET with instructor debriefing trainees (Source: Petans Training)

Figure 9.21 Helicopter Rescue Simulator (Source: Warsash Maritime Academy)

   b. Using EBS without windows. The next stage or partial capsizing is to use the EBS as learnt through EBS acquaintance exercise and combine the two drills.

   c. Using EBS with windows. In this run, HUET windows are put in place with trainees using their EBS when they exit the trainer.

5. Full HUET capsize: When trainees complete the partial HUET capsize drills, then they are taken through the three drills below where the HUET is capsized by 180° on each run.

   a. without EBS without windows;

   b. with EBS without windows;

   c. with EBS with windows.

Subsequent to the above exercises, the trainees inflate their life-jacket upon exit from the HUET and carry out survival exercises leading to rescue from the sea by winching.

6. Rescue by winching. This is usually the final exercise carried out at the end of the HUET training session. When all trainees are gathered in the helirafts, they are winched up on to a simulated helicopter platform as shown in Figure 9.21. This procedure simulates the Hi-Line technique described in Chapter 7.

Notes

1 CAA (2014) Civil Aviation Authority – Safety Review of Offshore Public Transport Helicopter Operations in Support of the Exploitation of Oil and Gas. CAP 1145. Available at www.caa.co.uk/ [Accessed 17.11.14].

2 HSE (2005) How Offshore Helicopter Travel is Regulated. HSE Publication HSG 219. Available at www.hse.gov.uk/pubns/indg219.pdf [Accessed 09.02.15].

3 HSE (2009) HSE/SRG Memorandum of Understanding, Issue 3 and Annex 4. Available at www.caa.co.uk/ [Accessed 09.02.15].

4 CAA (2014) Civil Aviation Authority – Safety Review of Offshore Public Transport Helicopter Operations in Support of the Exploitation of Oil and Gas. CAP 1145. Available at www.caa.co.uk/ [Accessed 10.02.15].

5 Agusta (2014) Description of the AB139 HUMS. Presentation on AW139 HUMS by AgustaWestland Co. Available at www.skybrary.aero/bookshelf/books/2637.pdf [Accessed 10.02.15].

6 NTIS (1989) US Congress, Office of Technology Assessment, Safer Skies with TCAS: Traffic Alert and Collision Avoidance System – A Special Report, OTA-SET-431 (Washington, DC: US Government Printing Office, February 1989); ICAO (2006) Airborne Collision Avoidance System (ACAS) Manual. Available at www.icao.int/ [Accessed 11.02.15]; FAA (2011) Introduction to TCAS II. US Department of Transportation, Federal Aviation Administration. Available at www.faa.gov/ [Accessed 11.02.15].

7 CAA (2010) Master Minimum Equipment Lists (MMEL) and Minimum Equipment Lists (MEL) CAA publication CAP 549. Available at www.skybrary.aero/bookshelf/books/822.pdf [Accessed 14.02.15].

8 ICAO (2010) Operation of Aircraft – Annex 6 to the Convention on International Civil Aviation. Available at www.icao.int [Accessed 14.02.15].

9 CAA (2014) Civil Aviation Authority, UK, How to Apply for the ATPL(H) Licence. Available at www.caa.co.uk/ [Accessed 14.02.15].

10 OGP (2013) OGP – Aircraft Management Guidelines. Available at www.ogp.org.uk/pubs/390.pdf [Accessed 14.02.15].

11 Available at www.ihst.org/LinkClick.aspx?fileticket=uhdMiyXCSCE= [Accessed 20.02.15].

12 HSSG (2014) HSSG Snapshot – EBS. Step Change in Safety. Available at www.oilc.org/hssg_snapshot_-_ebs-1.pdf [Accessed 12.02.15].

10

Asset Integrity

Offshore structures are placed in harsh environments where structural integrity is dependent upon the ability of the operators to maintain all hardware to a robust, efficient and sustainable standard to protect life, the environment and assets. Obviously, this will require great attention to detail to ensure a safe and incident free environment for the offshore workforce.

Asset integrity in essence is the safety of all components in an installation that ensure process safety through focusing on potential hazards that may lead to major accidents such as fire, explosion, collision with vessels or structural failures affecting the installation as a whole and therefore causing catastrophic consequences. HSE defines asset integrity as 'the ability of an asset to perform its required function effectively and efficiently whilst protecting health, safety and the environment'[1]. In comparison with occupational safety that encompasses personal safety, the scope of asset integrity is vast where the former may be incorporated within the latter. Asset integrity management is the identification and availability of resources including workforce, systems and processes to achieve asset integrity.

Whilst it is a known fact that in order to make profits, the facility must continue to produce the required product, it was only in 2004 when Step Change in Safety introduced 'asset integrity' as the third pillar of the temple model[2] shown in Figure 10.2 as a result of Oil and Gas UK establishing the Installation Integrity Work Group involving over 30 offshore operators and other bodies.

Figure 10.1 Surveyors inspecting assets must inspect all parts of an installation (Source: © Petrofac Training)

The resulting product of this exercise was an Asset Integrity Tool Kit which is available on the Step Change in Safety website at www.stepchangeinsafety.net.

The six stages of the lifecycle of any offshore asset are design, construction and hook-up, commissioning, operations, modifications and decommissioning. The main components of the integrity lifecycle for Safety Critical Elements (SCE) identified by the work group are as below:

1. Assurance and Verification
2. Assessment/Control and Monitoring
3. Competence
4. Planning
5. Maintenance
6. Quality and Audit.

10.1 Monitoring Assets

Two models i.e. the Swiss Cheese Model and the Bow-Tie Model, are described in the next sections; these can be used to monitor an asset's integrity throughout its lifecycle.

10.1.1 The Swiss Cheese Model

All offshore installations must have an Asset Integrity Management System (AIMS) that provides scope of an asset's outputs against design functions for the entire lifecycle of the project. Each stage of asset lifecycle such as design, operation, production, maintenance and decommissioning including assurance audits or inspections to

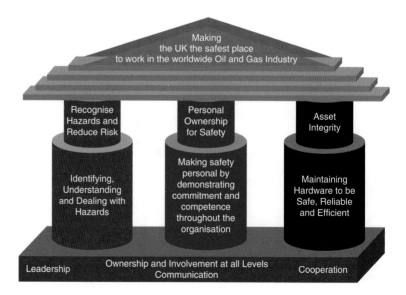

Figure 10.2 Three pillars of asset integrity (Source: adapted from Step Change in Safety)

Plant   Place   Competency System

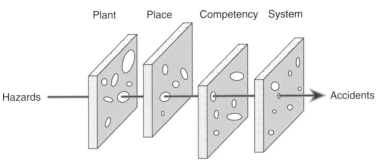

Hazards

Accidents

Figure 10.3 The Swiss Cheese model

monitor deliverables must be incorporated into AIMS. In order to achieve an effective AIMS, it must incorporate a series of 'barriers' to mitigate the consequences of a hazard realisation. A common model to show this principle is known as the Swiss Cheese Model, shown in Figure 10.3. The holes in the model shown are the weaknesses that will allow the hazard to escape through each barrier (cheese). If a number of holes in various barriers are aligned, then the hazard can escape through all barriers leading to an accident. These holes can be anywhere in the four given areas i.e. plant, place, competency or system as described in Chapter 2.

All components of an installation have a service life during which these will degrade with a potential to decrease output, which may eventually lead to failure unless adequately maintained or replaced.

Each component within any system whose failure can lead to an accident is referred to as a 'safety critical element' and therefore considered as a barrier to prevent an incident. This element can be any part that can be included in any of the four categories mentioned earlier i.e. plant, place, competency or system. The barrier may be just one or a combination of these components.

For example, a pressure relief valve may be a standalone component on a gas tank to release an over-pressure whereas the pressure monitors combined with 'trip' switch(es) to stop pumping when the desired pressure is reached is another barrier. Personnel ensuring correct functioning of these components are a third barrier. Each one of these components can be considered 'safety critical' as consistent failure in any component will eventually lead to an accident.

Maintenance activities should therefore ensure both the performance and integrity to ensure safeguards against component failure and thus maintain overall integrity of the asset. However, this can only be achieved through an understanding of the 'deterioration' process which will in turn lead to the development of AIMS that encapsulates top-down commitment from management including involvement of workers at all levels. A backlog of maintenance provides evidence of an incomplete system that will eventually lead to failure of barriers allowing hazards to develop into full-fledged disasters. A standard approach to manage risks associated with asset integrity is based on Plan, Do, Check, Act (PDCA) as described in Chapter 2.

10.1.2 The Bow-tie Model

This model provides an understanding of the linkage between various hazards leading to accidents as well as the methods to recover from identified consequences. Initially, the hazards are identified and then a bow-tie analysis is carried out to identify the risks, control measures, consequences of failure of control measures and procedures to recover from any undesired outcome. The benefit of using this technique is that the bow-tie provides a clear link between the hazards, control measures and consequences as well as areas where senior management must take proactive action to avoid the identified adverse outcome. The added benefit is that the bow-tie analysis is mostly carried out in pictorial format and presented to all stakeholders who can then input into it to refine it further.

10.1.3 Asset Integrity Barrier Examples

Asset integrity barriers can be divided into four categories:[3]

1. Prevention – this category includes primary means to contain hazards though process control and structural measures.
2. Detection – when the primary means fail, SCE in this category will come into play such as alarms for smoke, fire or gas leakage.
3. Control and Mitigation – upon detection of any hazards, control is required through systems such as fire suppression systems, blow down preventers, etc.
4. Emergency Response – When all three preceding barriers partially or fully fail, then the safe evacuation of personnel as well as procedures to reduce the damage are considered, which is based on means for escape or evacuation and methods to reduce the impact through ensuring adequate means of communication and emergency power supply.

All barrier categories require competent personnel who will ensure continuous availability of reliable barriers, which includes training of all personnel for the allocated tasks. Within the project

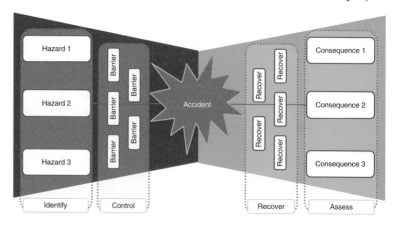

Figure 10.4 The Bow Tie model

specifications, performance standards based on functionality, availability and reliability are initially defined and subsequently checked through audits and reviews to ensure their continued adequacy and fitness for purpose for which they are deployed.

10.2 Offshore Installation Safety Case

The 6 July 1988 Piper Alpha disaster in the North Sea killing 167 workers and completely destroying the installation led to a formal inquiry headed by Lord Cullen. Consequently a report was produced that contained recommendations to shift legislation regime from prescriptive to goal-setting, concentrating primarily on Major Accident Hazard (MAH) management. Starting from early 1990s,

- Offshore Installations (Safety Case) Regulations (SCR) introduced major hazard assessment;
- Offshore Installations (Design and Construction) Regulations (DCR) introduced Safety Critical Elements (SCE).

In addition, KPI (Key Performance Indicator) arrangements were put in place as a measure to monitor and therefore improve performance.

A safety case requires the duty holder to demonstrate their ability and show resources to control risks of accidents effectively. HSE's offshore division has been instrumental in taking the Cullen Report recommendations forward by introducing KP1 and KP3 programmes which has now moved on to the KP4 programme. KP1 focused on the monitoring of hydrocarbon releases and the integrity of process plant. However, the ageing plant caused concern which led to the introduction of the KP3 programme in 2004 where the concept of asset integrity was widened to include existing elements of KP1 as well as the physical condition of plant.

After the end of the KP3 programme in 2007, 3 KPIs were introduced through HSE's KP3 programme.

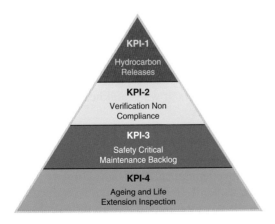

Figure 10.5 KPIs

In 2010, the KP4 – Ageing and Life Extension Inspection Programme was introduced due to the fact that ageing of offshore installations is a major concern. The programme requires operators to submit reports to HSE about the risks associated with ageing structures and measures being taken to mitigate these risks.

- KPI – 1: Hydrocarbon Releases
- KPI – 2: Verification Non-compliance
- KPI – 3: Safety Critical Maintenance Backlog
- KPI – 4: Ageing and Life Extension Inspection.

Mandated through legislation, HSE requires the reporting of certain parts of KPIs to monitor performance in vital areas whereas provision of certain additional information is kept voluntary to participating operators.

The areas covered in the KP4 programme are:

- Ageing – This is not the chronological 'age' of the installation or its components but as defined by HSE, ageing is the effect whereby a component suffers some form of material deterioration and damage with an increasing likelihood of failure over the lifetime. Factors such as corrosion, erosion or fatigue may have varied levels of deterioration impact on the installation if preventive measures have not been adequate during the life of the installation.
- Corrosion – This is a chemical process in which a metal is physically degraded through exposure to an environment that supports such chemical processes. There are two types of corrosion, both taking places on the surface of metals. Wet corrosion takes place due to presence of water or moisture whereas dry corrosion occurs at temperatures above 400°C due to the reaction of metal to the oxygen in the atmosphere.
- Erosion is a form of corrosion but it is not a chemical reaction and takes place much more quickly than normal corrosion. This usually occurs due to particles suspended in flowing liquid or gas which causes abrasion, removes the protective layer from the metal surface and assists in forming an oxide layer, thus causing erosion. The faster the rate of flow, the quicker can be the rate of this type of corrosion. Exposed surfaces of structures may suffer from a combination of erosion and corrosion due to the effect of wind and seawater spray. Most obvious places where erosion is common are corners of walls/bulkheads, exposed beam webs, brackets and flanges.
- Fatigue – Offshore structures are prone to cyclic stresses due to vibration, impact of sea/swell waves and wind. Monitoring the impact of fatigue on offshore structures can be very challenging due to their complicated design. However, whilst detailed monitoring can only be carried out by trained engineers, all workers must pay attention to any abnormal observations such

as cracks in metal work and report them immediately though their supervisors.

- Physical damage to offshore structures can result for several reasons. For example, concrete foundations of the structures may suffer from cracks resulting from freezing of water in pores, marine growth, and crystallisation of salt. Other physical damage may occur to primary barriers such as protective coatings which naturally degrade over time but poor workmanship during the first application may result in rapid degradation.

- Weathering – Seawater contains chlorides, sulphates, sodium, magnesium and other types of salts. The quantity of these salts vary from one place to another, hence no standard rules can be applied to protect the structures from adverse impact from them. However, the most important characteristics of seawater which are considered during the design stage are its pH value, oxygen content, salinity and temperature because their values have a significant impact on their chemical reaction with steel work. Salinity in particular plays a significant role in determining the rate of corrosion in any steel structure because conductivity of seawater depends upon salinity. The higher the salinity, the quicker currents can flow though seawater and therefore the higher is the rate of corrosion.

- Expansion/Contraction – Offshore structures are subjected to a large number of operational loads including weight, internal or external pressure as well as changes in temperatures. Under any load, the structure is likely to contract whereas when the load is removed, it will expand. A change in temperature brings about a similar change for expansion at higher temperatures and contraction at lower temperatures. Consequently, the structures suffer from a continuous motion within them that may lead to structural failure.

- Subsidence – Offshore platforms are likely to undergo subsidence due to production activities or other movements of the seabed. During the extraction of hydrocarbons, the seabed is expected to naturally suffer from compaction. Various technologies are used to monitor subsidence using various benchmarks at known locations. The position of various objects in relation to those benchmarks is then monitored.

- Process Safety – The industry in the UK has successfully developed and implemented a process safety culture which involves workers at all levels. Various tools have been developed to assess safety culture to ensure monitoring of the implementation of all required procedures and practices. Details of safety culture are discussed in Chapter 2.

- Electrical, Control and Instrumentation (E/C&I) Systems include relays, switches, sensors, transmitters and pumps, etc. which form part of the emergency shutdown systems alarms and are combined with other systems to ensure overall safety of

Figure 10.6 Asset integrity monitoring involves close inspection of all components of an installation

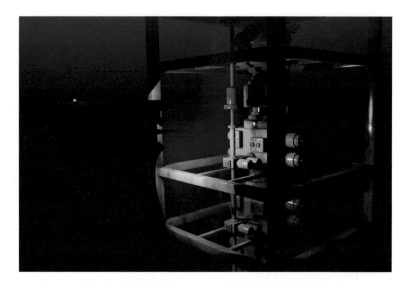

the installation. Whilst in most of these systems, there may not be any moving components, they suffer from degradation with time in the same way as other equipment and hence the same general rules apply. Specific guidance has been written for the industry to understand and mitigate the issues caused by various aforementioned reasons. One such guide is the 'E/C&I Plant Ageing: A Technical Guide for Specialists Managing Ageing E/C&I Plant'.

• Human factor is almost always contributory to the majority of accidents. Improvement in the competency of the personnel involved eventually leads to ensuring safety at all levels. All stages of an offshore project require differing levels of competencies for the workers involved. Any gaps must be identified prior to assigning workers to the tasks and training organised as required.

10.3 Divisions of Asset Integrity

Asset integrity is divided into four areas:

1. Design Integrity: Ensuring installation including its facilities' design complies with industry standards as well as operational requirements.
2. Technical Integrity: Ensuring inspection and maintenance processes ensure continuous and safe availability of systems including data to analyse information gathered for decision making. This includes prioritisation of tasks that are safety critical.
3. Leadership Integrity: Top-down commitment from senior management at all stages of design, operations and maintenance of an installation.

4. Operating Integrity: Availability of sufficient competent work-force to ensure optimal safe use of installation throughout its lifecycle.

As per HSE's KP3 programme,[4] key findings of the asset integrity were based on the following areas:

1. Asset integrity/process safety management
2. Physical state of plant
3. Safety critical systems
4. Leadership
5. The engineering function
6. Corporate and cross-industry learning and communication
7. Human resources and competence
8. Safety culture
9. Workforce involvement in controlling major accident hazards
10. Existing mechanisms for workforce involvement.

Figure 10.7 Asset integrity divisions

10.4 Employer and Employee Responsibilities for Asset Integrity

Both the employer and employees have responsibilities to look after their assets to ensure continuous productivity as well as the safety of life and environment. Therefore they must ensure:

1. compliance with legislation and performance standards for all processes and procedures;
2. fitness of physical assets for their design intent for the full life-cycle of the asset;
3. review of maintenance procedures to ensure ongoing safe operations;
4. compliance with manufacturer's instructions for safe operations of equipment within specified limits;
5. understanding of procedures to operate equipment including its limitations;
6. corrosion control through internal and external corrosion control programmes;
7. inspection, testing and maintenance to the highest possible standards of safety critical items;
8. availability of equipment in a fully fit state for its intended purpose;
9. availability of third party verification and certification where required;
10. adequate competent workers to perform their job roles safely and developing an understanding to link their role with asset integrity;

11. compliance with installation procedures at all times;
12. all personnel intervene when they observe lack of compliance with any agreed procedures or workers taking shortcuts with a potential to cause undue risk;
13. all personnel follow the installation's procedures to report unsafe conditions to their supervisors;
14. the workforce develops a culture of accountability for their acts and omissions;
15. relevant policies are continually updated in pace with any changes in processes, procedures or equipment;
16. support for continuous development of the management systems pertaining to asset integrity and process safety;
17. regular audits, reviews or independent verifications are conducted to warrant compliance at all times.

10.5 Decommissioning

Decommissioning does not always mean complete removal of the installation. According to an OGP report,[5] there are four decommissioning options:

1. reuse at existing location for any commercial or research activities or to capture or store energy related products;
2. full removal to use at another location, deconstruction, recycling and disposal onshore or offshore; or demolition offshore and then transportation onshore for disposal;
3. partial removal to facilitate transportation;
4. leave in place but remove topside to make the area safe for surface navigation.

The areas of the seas around the world are filled with a large number of offshore structures. In the North Sea alone, there are some 600 platforms of the following four types:

- Less than 4000 tonnes steel platforms – 377
- Above 4000 tonnes steel platforms – 106
- Concrete Gravity Based Structures (CGBS) – 24
- Floating platforms – 45.

It has been agreed by the industry and the participating governments that all offshore structures in the OSPAR region of the North East Atlantic must be removed after completion of duties as given in the 1992 Convention for the Protection of the Marine Environment in the North Atlantic. Only one exception was allowed for the concrete platform because of complexities of the operations to remove them. However, topsides of the Concrete Gravity Based Structures (CGBS) must be removed but concrete can be left in situ because of the low environmental impact. The industry has estimated a cost of around £1 billion to remove a complete platform successfully. From an offshore operator's perspective, an offshore structure will be decommissioned if it requires more expenditure in comparison with the possible outputs therefore increasing the production costs. This may be due to the age of the structures requiring more maintenance, long-term risk to life or environment or even the possibility of reusing the structure or its components elsewhere.

The International Maritime Organization (IMO), however, recommends that:

- All structures weighing less than 4000 tonnes in water depths of less than 100 m must be completely removed.
- Structures in deeper waters can be partially removed provided a minimum of 55 m clearance above the structure is maintained.
- The position of the partially removed structures must be shown on the navigational charts with any remains marked with appropriate navigational aids to avoid collision with surface navigation vessels.

Decommissioning can be a long and tedious process that requires perhaps more planning than the initial design and commissioning. Whilst in theory, it can be a 'reverse engineering' project, in practice, the project managers must be prepared for the worst. An example is the complications in decommissioning of Total's TCP2 module support frame (MSF) platform in the North Sea.[6] Features of this undertaking include:

- Weight of the module – 8730 tonnes
- Weight of the module, trailers and fittings to bring the module ashore – 11,100 tonnes
- Total time spent on de-pollution measures – approximately 4 months
- Deconstruction time – approximately 6 months.

Notes

1 HSE (2007) Key Programme 3 Asset Integrity Programme. A report by the Offshore Division of HSE's Hazardous Installations Directorate [Online] www.hse.gov.uk/offshore/kp3.pdf [Accessed 22.11.14].
2 HSE (2007) Key Programme 3 Asset Integrity Programme. A report by the Offshore Division of HSE's Hazardous Installations Directorate [Online] www.hse.gov.uk/offshore/kp3.pdf [Accessed 22.11.14].
3 OGP (2008) Asset Integrity – the Key to Managing Major Incident Risks. International Association of Oil and Gas Producers Report No. 415.

4 HSE (2009) Key Programme 3 - Asset Integrity: A Review of Industry's Progress. A report by the Offshore Division of HSE's Hazardous Installations Directorate. [Online] www.hse.gov.uk/offshore/kp3review.pdf [Accessed 08.02.15].
5 OGP (2012) Decommissioning of Offshore Concrete Gravity Based Structures (CGBS) in the OSPAR Maritime Area/Other Global Regions. Report No. 484.
6 Total (2011) Frigg Field Cessation Plan Close Out Report. Stavanger, Norway.

11

Training for the Offshore Oil and Gas Industry

11.1 Introduction to OPITO

In the UK in particular and across other areas of the world in general, the workforce in the offshore oil and gas industry trains to standards developed by the industry under the umbrella of OPITO. Today, some 250,000 offshore workers undergo training designed and approved according to the OPITO standards.

Originally set up as PITB (Petroleum Industry Training Board) in 1977, this was split in 1982 into OPITB (Offshore Petroleum Industry Training Board) and PTF (Petroleum Training Federation). In 1991, OPITO was set up as Offshore Petroleum Industry Training Organisation under a government scheme to establish National Training Organisations (NTO). The acronym OPITO became so known to the industry that a decision was made to keep the name of organisation OPITO as a proper noun.

Currently, OPITO has offices in Aberdeen (head office) with branches in Kuala Lumpur (Malaysia), Dubai (UAE) and Houston (USA). OPITO operates with a mission[1] 'to support a competent and safe workforce and to ensure that quality, innovation and partnership underpin everything that we do'.

Since its inception, OPITO has developed a large number of training standards for UKCS, many of which have been accepted in their entirety in many other parts of the world. Some of these standards have been adopted with amendments to suit local variations under OPITO's globalisation scheme.

OPITO's training and competency standards are divided into the following areas:

1. Emergency and critical response standards
2. Specialised emergency response
3. Industry training and competency standards
4. UK national occupational standards.

The information below is current at the time of writing this book and as available from OPITO Standards' Library. The most current requirements must be checked from OPITO's website at www. OPITO.com.

11.2 Emergency and Critical Response Standards

Training standards in this category include subjects covering survival and emergency response. This category is further subdivided into the following areas.

11.2.1 Basic Emergency Response

All workers employed in the UKCS are required by the duty holder to complete some of these courses: Basic H_2S Training, Basic Offshore Safety Induction and Emergency Training (BOSIET), Further Offshore Emergency Training (FOET), Escape Chute Training (ECT), Helicopter Underwater Emergency Training (HUET) and Travel Safely by Boat.

11.2.1.1 Basic H_2S Training

H_2S or hydrogen sulphide gas is one of the most lethal chemical gases that can be found in workplaces in the offshore industry. The dangers associated with this gas become more significant due to its characteristics of being colourless, flammable, heavier than air and highly toxic at even extremely low concentrations. At low concentrations, it smells like rotten egg but once inhaled, it causes a loss of sensitivity to smell.

In any workplace where the concentration of H_2S is likely to be above 10 parts per million (ppm), workers should be provided with training to recognise its presence using H_2S detection equipment which will determine the presence of H_2S and also assist in establishing whether a worker is likely to be exposed above the Occupational Exposure Limit (OEL).

A brief summary of OPITO's H_2S course is given here for readers.

Target group	All personnel likely to be exposed to H_2S at their workplace
Pre-requisite training	None
Medical fitness	Valid offshore medical certificate or medical screening by the training provider
Training objectives	To provide knowledge of H_2S properties and hazards associated with exposure to it and response to exposure incidents

Learning outcomes	The training includes all necessary topics to cover formation of H_2S, its properties, OEL, detection and respiratory equipment for areas where personnel may be exposed to H_2S. In addition to covering the principles, the training also includes practical exercises for use of H_2S detection equipment, response to alarm and use of Escape Breathing Apparatus (EBA).
Assessment	Demonstration of equipment Written examination
Duration	4 hours
Validity of certificate	2 years
Refresher training	Basic H_2S training (Same courses as initial training)

11.2.1.2 Basic Onshore Emergency Response (BOER)

This training standard was introduced in 2012 to make training similar to BOSIET available for the onshore oil and gas industry and therefore covers similar topics. However, because of the difference in the location of workplaces, the offshore travel and survival elements are not covered in the BOER course.

Target group	All personnel working in the onshore oil and gas industry
Pre-requisite training	None
Medical fitness	Valid offshore medical certificate or medical screening by the training provider
Training objectives	To provide necessary knowledge of emergency response to onshore oil and gas industry workers
Learning outcomes	Knowledge of hazards in the onshore oil and gas industry that can lead to fire, emergency response, use of fire fighting equipment, and self-rescue and emergency first aid. The course is split into 40% classroom theoretical training and 60% practical training.
Assessment	Observation of demonstration in practical exercises, end of course open book written assessment
Duration	8 hours
Validity of certificate	4 years
Refresher training	Further Onshore Emergency Response (FOER)

11.2.1.3 BOSIET – Basic Offshore Safety Induction and Emergency Training

One of the most critical factors in reaching the workplace in the offshore oil and gas industry is travelling to and from the installation. As discussed in Chapter 9, personnel may travel by helicopter or by boat but with helicopter travel being more critical, survival and actions in an emergency related to travel by helicopter are covered in this course. In addition, the course also covers induction to hazards in the workplace, HUET, use of TEMPSC as a passenger, escape from smoke, and basic fire fighting and emergency first aid are also covered to provide workers with a basic knowledge for ensuring safety and self-survival.

Target group	All personnel working in the offshore oil and gas industry
Pre-requisite training	None
Medical fitness	Valid offshore medical certificate or medical screening by the training provider
Training objectives	To provide necessary knowledge of emergency response in relation to applicable legislation to offshore oil and gas industry workers
Learning outcomes	Knowledge of hazards in the offshore oil and gas industry, response to emergencies, reporting of incidents/accidents, use of prescribed medicine, use of fire fighting equipment and self-rescue from smoke filled areas, use of PPE when travelling and working offshore by helicopters, sea survival and emergency first aid
Assessment	Observation of demonstration in practical exercises, open book written assessment
Duration	20 hours 25 minutes spread over 3 days
Validity of certificate	4 years
Refresher training	Further Offshore Emergency Training (FOET)

11.2.1.4 Compressed Air Emergency Breathing System (CA-EBS) Initial Deployment Training

This training has been offered since July 2014 and was introduced as a result of the CAA report about a helicopter ditching in the North Sea in 2013. The report recommended the following.

With effect from 1 June 2014, the CAA prohibits the occupation of passenger seats not adjacent to push-out window emergency exits during offshore helicopter operations, except in response to an offshore emergency, unless the consequences of capsize are mitigated by at least one of the following:

1. all passengers on offshore flights wearing Emergency Breathing Systems that meet Category 'A' of the specification detailed in CAP 1034 in order to increase underwater survival time;
2. fitting the side-floating helicopter system in order to remove the time pressure to escape.

Further details of EBS systems can be found in Chapter 9 in section 'Emergency Breathing Systems (EBS)'.

Target group	All personnel travelling by helicopter and likely to use CA-EBS
Pre-requisites	In date BOSIET, FOET or HUET certificate
Medical fitness	Valid offshore medical certificate or medical screening by the training provider

Objectives	To provide knowledge of the hazards associated with use of CA-EBS and response to the need for its emergency deployment
Learning outcomes	Understand the principles of CA-EBS and difference between re-breather and CA-EBS. Demonstrate inspection of CA-EBS and lifejacket and deployment of CA-EBS in dry environment
Assessment	Observation of demonstration in practical exercises and a written assessment
Duration	1 hour 30 minutes
Validity	No expiry but duty holder may impose refresher training in alignment with BOSIET/FOET courses
Refresher training	Not applicable

11.2.1.5 Emergency Breathing System (delivered in conjunction with T-BOSIET and T-HUET) – EBS-T-BOSIET/EBS-T-HUET/T-FOET

Target group	**All personnel travelling offshore by helicopter in warm water tropical environment issued with EBS**
Pre-requisite training	In date T-BOSIET, T-FOET, BOSIET, FOET certificate
Medical fitness	Valid offshore medical certificate or medical screening by the training provider
Training objectives	To provide knowledge of use of EBS for attendees undertaking T-BOSIET/T-FOET/T-HUET courses
Learning outcomes	Demonstrate use of EBS deployment during practical helicopter underwater escape training
Assessment	Observation of demonstration in practical exercises
Duration	6 hours 30 minutes with 10% theory and 90% practical training (when delivered in conjunction with T-BOSIET/T-HUET) 4 hours (when delivered in conjunction with T-FOET)
Validity	4 years
Refresher training	Not applicable. To be carried out in conjunction with T-FOET or T-HUET courses.

11.2.1.6 Escape Chute Training (ECT)

Escape chutes have been used in the offshore industry as tertiary means of escape from offshore installations. Until 2012–2013, OPITO did not have any approved standard for ECT training whilst other countries in the North Sea sector required it. As many of the British workers were employed in EU sectors outside of the UKCS, they used to undertake non-OPITO approved training recognised by the duty holders in those regions.

Target group	**Personnel employed in workplaces where the duty holder requires them to undertake ECT training in addition to BOSIET, FOET, T-BOSIET or T-FOET courses**
Pre-requisites	None
Medical fitness	Valid offshore medical certificate or medical screening by the training provider
Training objectives	To give delegates a knowledge of the escape chutes as a means of tertiary escape from offshore installations and/or vessels
Learning outcomes	Understand the types of escape chutes found in the offshore industry and demonstrate use of escape chute to escape from the installation
Assessment	Observation of demonstration in practical exercises
Duration	1 hour 50 minutes
Validity	4 years
Refresher training	ECTR

11.2.1.7 Escape Chute Training Refresher (ECTR)

Target group	**See ECT above**
Pre-requisite training	None
Medical fitness	See ECT above
Training objectives	See ECT above
Learning outcomes	See ECT above
Assessment	See ECT above
Duration	1 hour 50 minutes
Validity of certificate	4 years
Refresher training	ECTR

11.2.1.8 Further Offshore Emergency Training (FOET)

Target group	**All personnel working in the offshore oil and gas industry**
Pre-requisite training	In date BOSIET, T-BOSIET, FOET or T-FOET certificate
Medical fitness	Valid offshore medical certificate or medical screening by the training provider
Training objectives	To refresh knowledge of emergency response gained in BOSIET/T-BOSIET course which could not be practised at workplace
Learning outcomes	Knowledge of preparation for dealing with helicopter emergencies, fire fighting and self-rescue and basic first aid
Assessment	Observation of demonstration in practical exercises
Duration	8 hours

Validity of certificate	4 years
Refresher training	Further Offshore Emergency Training (FOET)

11.2.1.9 Further Onshore Emergency Response (FOER)

Target group	**All personnel working in the onshore oil and gas industry who have previously undertaken BOER course**
Pre-requisite training	In date BOER or FOER certificate
Medical fitness	Valid offshore medical certificate or medical screening by the training provider
Training objectives	To refresh necessary knowledge of emergency response to workers who gained initial BOER training
Learning outcomes	Knowledge of mustering procedures, use of portable fire extinguishers, self-rescue using smoke hoods and emergency first aid
Assessment	Observation of demonstrations in practical exercises
Duration	6 hours
Validity of certificate	4 years
Refresher training	Further Onshore Emergency Response (FOER)

11.2.1.10 Helicopter Underwater Emergency Training (HUET)

Target group	**Personnel travelling offshore by helicopter**
Pre-requisite training	None
Medical fitness	Valid offshore medical certificate or medical screening by the training provider
Training objectives	To provide the knowledge of pre-flight and in-flight procedures and prepare travellers for responding to helicopter emergencies when travelling offshore
Learning outcomes	Use Emergency Breathing System (EBS), actions following a helicopter ditching
Assessment	Observation of demonstrations in practical exercises
Duration	6 hours 40 minutes
Validity of certificate	4 years
Refresher training	Retake HUET initial course

11.2.1.11 T-BOSIET – Tropical Basic Offshore Safety Induction and Emergency Training

Target group	**Personnel travelling offshore by helicopter in tropical areas**
Pre-requisite training	None

Medical fitness	Valid offshore medical certificate or medical screening by the training provider
Training objectives	To provide necessary knowledge related to safety issues and emergency response as required by legislation applicable to offshore oil and gas industry workers employed in tropical areas
Learning outcomes	Same as BOSIET except no EBS training is required. Knowledge of hazards in the offshore oil and gas industry, response to emergencies such as fire, medical emergencies, reporting of incidents/accidents, use of prescribed medicine, use of fire fighting equipment and self-rescue from smoke filled areas, use of PPE when travelling and working offshore by helicopters, sea survival and emergency first aid.
Assessment	Observation of demonstration in practical exercises, open book written assessment
Duration	18 hours 05 minutes with 40% theory and 60% practical
Validity of certificate	4 years
Refresher training	Tropical Further Offshore Emergency Training (T-FOET)

11.2.1.12 T-BOSIET/FOET to BOSIET/FOET – Bridging Elements Training

Target group	**Personnel who have undertaken OPITO T-BOSIET/T-FOET and wish to upgrade their qualification to BOSIET/FOET**
Pre-requisite training	In date T-BOSIET or T-FOET certificate
Medical fitness	Valid offshore medical certificate or medical screening by the training provider
Training objectives	To undertake skills and knowledge of the use of aviation transit suits and EBS for travel by helicopter
Learning outcomes	Skills to use aviation transit suit and EBS
Assessment	Observation of demonstration in practical exercises
Duration	
Validity of certificate	
Refresher training	Further Offshore Emergency Training (FOET)

11.2.1.13 T-FOET – Tropical Further Offshore Emergency Training

Target group	**All personnel working in the offshore oil and gas industry in tropical areas**
Pre-requisite training	In date BOSIET, T-BOSIET, FOET or T-FOET certificate
Medical fitness	Valid offshore medical certificate or medical screening by the training provider
Training objectives	To refresh knowledge of emergency response gained in BOSIET/T-BOSIET course which couldn't have been practised at workplace
Learning outcomes	Knowledge of dealing with helicopter emergencies, fire fighting and self-rescue and basic first aid

Assessment	Observation of demonstration in practical exercises
Duration	7 hours 05 minutes
Validity of certificate	4 years
Refresher training	Tropical Further Offshore Emergency Training (FOET)

11.2.1.14 T-HUET – Tropical Helicopter Underwater Escape Training

Target group	**Personnel travelling offshore by helicopter in tropical areas**
Pre-requisite training	None
Medical fitness	Valid offshore medical certificate or medical screening by the training provider
Training objectives	To provide the knowledge of pre-flight and in-flight procedures and prepare travellers for responding to helicopter emergencies when travelling offshore
Learning outcomes	Actions following a helicopter ditching including survival at sea
Assessment	Observation of demonstrations in practical exercises
Duration	8 hours
Validity of certificate	4 years
Refresher training	Retake T-HUET initial course

11.2.1.15 Travel Safely by Boat – Initial Training (TsbB)

Target group	**New entrants to the offshore oil and gas industry travelling offshore by boat**
Pre-requisite training	None. Holders of BOSIET/T-BOSIET or FOET/T-FOET are not required to undertake sea survival exercises involving liferafts
Medical fitness	Valid offshore medical certificate or medical screening by the training provider
Training objectives	To provide knowledge of safety rules associated with travelling by boat, actions in an emergency, use of personal safety equipment and transfer from boat
Learning outcomes	Donning of floatation devices when transferring from boat, actions during emergencies on boats, precautions during rope transfer, survival techniques including boarding a liferaft
Assessment	Observation of demonstrations in practical exercises
Duration	2 hours
Validity of certificate	4 years
Refresher training	Retake T-HUET initial course

11.2.1.16 Travel Safely by Boat – Further Training (FTsbB)

Target group	Personnel travelling offshore by boat
Pre-requisite training	In date Travel Safely by Boat (TsbB) Initial Training Certificate
Medical fitness	Valid offshore medical certificate or medical screening by the training provider
Training objectives	To refresh knowledge and skills gained in the initial TsbB training
Learning outcomes	Safety during embarkation/disembarkation to/from boat, donning of appropriate floatation devices, precautions during rope transfer, actions during an emergency, survival at sea including boarding a liferaft
Assessment	Observation of demonstrations in practical exercises
Duration	1 hour
Validity of certificate	4 years
Refresher training	Retake FTsbB initial course

11.3 Specialist Emergency Response

11.3.1 Command and Control for ERRV Masters and Mates

Target group	Masters, Mates (chief/second/third) who are likely to be in command of an ERRV operating offshore
Pre-requisite training	Deck officer qualification plus OPITO Initial Shipboard Operations Certificate
Medical fitness	Valid offshore medical certificate or medical screening by the training provider
Training objectives	To gain skills to successfully command an ERRV in emergency situations specific to the offshore oil and gas industry
Learning outcomes	Maintain vessel in a state of readiness, obtain information to analyse and assess situations, maximum utilisation of resources to take effective actions, maintain effective communications, delegation of authority and monitoring progress, self and team performance management, stress management to achieve the planned objectives and record keeping
Assessment	Observation of demonstrations in practical exercises
Duration	35 hours (35% theory, 65% practical)
Validity of certificate	Formal assessment through OODTP programme every 3 years OODTP – Ongoing On-board Development and Training Programme for ERRV Masters and Crews
Refresher training	See validity of certificate above

11.3.2 Control Room Operator Emergency Response Standard

Target group	**Personnel likely to be employed as a Control Room Operator (CRO)**
Pre-requisite training	BOSIET/FOET. Attendees are recommended to have completed Initial Introduction to Control Room Emergency Role and Responsibilities course
Medical fitness	Valid offshore medical certificate
Training objectives	To gain knowledge and skills required to control emergencies and critical situations
Learning outcomes	Maintain equipment in a state of readiness as required for the working environment, identify and control critical situations, coordinate response to emergencies by giving directions to personnel
Assessment	First element of this course needs to be completed prior to the attended elements 2 and 3 at a training provider. Further assessment will be based on process knowledge and three simulations to demonstrate competency in emergency response
Duration	Varies, dependent upon requirements by the duty holder
Validity of certificate	As per OPITO standard
Refresher training	Participation in at least one offshore drill/simulated emergency per year

11.3.3 Offshore Lifeboat Coxswain Free-Fall

Target group	**Personnel likely to be tasked the role of a coxswain for a free-fall lifeboat**
Pre-requisite training	BOSIET/FOET
Medical fitness	Valid offshore medical certificate or medical screening by the training provider
Training objectives	To gain knowledge and skills required to safely launch and recover the free-fall lifeboats
Learning outcomes	Role of coxswain in launch/recovery of a free-fall lifeboat, recovering survivors from water, steering the boat to a safe location using compass, preparing the lifeboat for launch, towing and pacing, use of means of communication, emergency equipment and pyrotechnics
Assessment	Observation of demonstrations in practical exercises
Duration	4 days
Validity of certificate	2 years
Refresher training	Offshore Lifeboat Coxswain Free-fall – Further Practice

11.3.4 Offshore Lifeboat Coxswain Free-fall – Further Practice

Target group	**Personnel holding an offshore lifeboat coxswain free-fall certificate**
Pre-requisite training	BOSIET/FOET and offshore lifeboat coxswain certificate
Medical fitness	Valid offshore medical certificate or medical screening by the training provider

Training objectives	To practise the skills for safe launch and recovery of free-fall lifeboats
Learning outcomes	Refreshing knowledge on the role of coxswain in launch/recovery of a free-fall lifeboat, recovering survivors from water, steering the boat to a safe location using compass, preparing the lifeboat for launch, towing and pacing, use of means of communication
Assessment	Observation of demonstrations in practical exercises
Duration	1.5 days
Validity of certificate	2 years
Refresher training	Offshore Lifeboat Coxswain Free-fall – Further Practice

11.3.5 Offshore Lifeboat Coxswain Twin Fall

Target group	**Personnel requiring initial training for becoming an offshore lifeboat coxswain**
Pre-requisite training	BOSIET/FOET and offshore lifeboat coxswain certificate
Medical fitness	Valid offshore medical certificate or medical screening by the training provider
Training objectives	To practise the skills for safe launch and recovery of the twinfall TEMPSC lifeboats
Learning outcomes	Role of coxswain in launch/recovery of a twinfall lifeboat, recovering survivors from water, steering the boat to a safe location using compass, preparing the lifeboat for launch, towing and pacing, use of means of communication, emergency equipment and pyrotechnics
Assessment	Observation of demonstrations in practical exercises
Duration	4 days
Validity of certificate	2 years
Refresher training	Offshore Lifeboat Coxswain Twin Fall – Further Practice

11.3.6 Offshore Lifeboat Coxswain Twin Fall – Further Practice

Target group	**Personnel holding an offshore lifeboat coxswain twinfall certificate**
Pre-requisite training	BOSIET/FOET and offshore lifeboat coxswain twinfall certificate
Medical fitness	Valid offshore medical certificate or medical screening by the training provider
Training objectives	To practise the skills for safe launch and recovery of the twinfall TEMPSC lifeboats
Learning outcomes	Refreshing knowledge on the role of coxswain in launch/recovery of a twinfall lifeboat, recovering survivors from water, steering the boat to a safe location using compass, preparing the lifeboat for launch, towing and pacing, use of means of communication
Assessment	Observation of demonstrations in practical exercises
Duration	1.5 days
Validity of certificate	2 years
Refresher training	Offshore Lifeboat Coxswain – Further Practice

11.3.7 Coxswain Supplementary Fall Training

Target group	**Personnel holding an offshore lifeboat coxswain in single fall lifeboats**
Pre-requisite training	BOSIET/FOET and offshore lifeboat coxswain certificate
Medical fitness	Valid offshore medical certificate or medical screening by the training provider
Training objectives	To practise the skills for safe launch and recovery of the twinfall TEMPSC lifeboats
Learning outcomes	Gaining knowledge on the role of coxswain in launch/recovery of a twinfall lifeboat, preparing the lifeboat for launch, use of means of communication
Assessment	Observation of demonstrations in practical exercises
Duration	4 hours
Validity of certificate	2 years
Refresher training	Offshore Lifeboat Coxswain – Further Practice

11.3.8 ERRV Crew Advanced Medical Aid

Target group	**Personnel requiring advanced medical aid training for ERRVs**
Pre-requisite training	OPITO Initial Training in Shipboard Operations
Medical fitness	Medical screening by the training provider
Training objectives	To gain skills required to become an advanced medical aider on ERRVs
Learning outcomes	Role of an ERRV advanced medical aider, maintaining a state of readiness to deal with emergencies, preparations to receive and treat casualties
Assessment	Observation of demonstrations in practical exercises
Duration	5 days
Validity of certificate	2 years
Refresher training	Advanced Medical Aid Further Training

11.3.9 ERRV Crew Advanced Medical Aid Further Training

Target group	**Personnel holding an advanced medical aid training for ERRVs**
Pre-requisite training	ERRV Crew Advanced Medical Aid Training
Medical fitness	Medical screening by the training provider
Training objectives	To maintain skills required to be an advanced medical aider on ERRVs
Learning outcomes	Role of an ERRV advanced medical aider, maintaining a state of readiness to deal with emergencies, preparations to receive and treat casualties
Assessment	Observation of demonstrations in practical exercises

Duration	2 days
Validity of certificate	2 years
Refresher training	Advanced Medical Aid Further Training

11.3.10 ERRV Crew Fast Rescue Craft Boatman

Target group	**Personnel employed on ERRVs and required to operate FRCs**
Pre-requisite training	Initial Training and Shipboard Operations plus three months sea time
Medical fitness	Medical screening by the training provider
Training objectives	To gain skills and knowledge required to prepare ERRV crew for assisting in launch/recovery and operation of FRCs
Learning outcomes	Assisting in preparation, launch, recovery of FRC, recovery and treatment of casualties, using communication equipment and boat handling including towing operations
Assessment	Observation of demonstrations in practical exercises
Duration	2 days
Validity of certificate	3 years
Refresher training	Revalidated through ongoing training programme on-board ERRVs

11.3.11 ERRV Crew Fast Rescue Craft Coxswain

Target group	**Personnel employed on ERRVs and required to be the coxswains of FRCs**
Pre-requisite training	Initial Training and Shipboard Operations and an ERRV Crew Fast Rescue Craft Boatman certificate
Medical fitness	Medical screening by the training provider
Training objectives	To gain skills and knowledge required to prepare for launch/recovery and operation of FRCs
Learning outcomes	Preparation, launch, recovery of FRC, recovery and treatment of casualties, using communication equipment and boat handling including towing operations, righting a capsized FRC
Assessment	Observation of demonstrations in practical exercises
Duration	4.5 days
Validity of certificate	3 years
Refresher training	Revalidated through ongoing training programme on-board ERRVs

11.3.12 ERRV Crew Daughter Craft Coxswain

Target group	**Personnel employed on ERRVs and required to be the coxswains of daughter craft carried on-board**
Pre-requisite training	ERRV Fast Rescue Craft Coxswain plus three month sea time as FRC coxswain
Medical fitness	Medical screening by the training provider
Training objectives	To gain skills and knowledge required to prepare for launch/recovery and operation of daughter craft
Learning outcomes	Preparation, launch, recovery and handling of daughter craft, knowledge of the international regulations for prevention of collision at sea, chartwork, buoyage, chart symbols and abbreviations, use of radar, GPS and VHF radio
Assessment	Observation of demonstrations in practical exercises
Duration	5 days
Validity of certificate	3 years
Refresher training	Revalidated through ongoing training programme on-board ERRVs

11.3.13 ERRV Crew Initial Training Shipboard Operations

Target group	**All crew members employed on ERRVs operating offshore**
Pre-requisite training	Personnel completing this training must join an ERRV as operational crew within 12 months of training completion, otherwise they are required to repeat this initial training programme
Medical fitness	Valid ENG1/offshore medical certificate or medical screening by the training provider
Training objectives	To gain skills to successfully command an ERRV in emergency situations specific to the offshore oil and gas industry
Learning outcomes	Use personal protective equipment, respond to emergencies on-board, assist in preparation of FRC, use of casualty handling devices, recover and care for casualties from sea
Assessment	Observation of demonstrations in practical exercises plus an oral/practical simulated exercise observed by an independent assessor
Duration	35 hours (35% theory, 65% practical)
Validity of certificate	Formal assessment through OODTP programme every 3 years OODTP – Ongoing On-board Development and Training Programme for ERRV Masters and Crews
Refresher training	See validity of certificate above

11.3.14 Offshore Helideck Assistant (HDA) Initial Training

OPITO HDA training is divided into five phases as given below:

1. HDA initial training
2. HERTM training
3. HDA workplace experience and competence assessment
4. HDA further training
5. HDA workplace drills and competence assessment – workplace assessment and reassessments are conducted on-board the offshore installation or the vessel as it requires site specific training and assessment. Details of the units assessed offshore are given in the OPITO HDA standard. This training consists of the following units:

   Unit 1: Maintaining a State of Readiness
   Unit 2: Preparations for Helicopter Landing and Departure
   Unit 3: Cargo Handling
   Unit 4: Refuelling the Helicopter
   Unit 5: Respond to Alarms and Emergencies.

A summary of the HDA initial training is given below whilst other shore based courses are described in the next subsections.

Target group	**Personnel tasked to assist on helideck during helicopter operations**
Pre-requisite training	None
Medical fitness	Offshore medical certificate or medical screening by the training provider
Training objectives	To gain knowledge and skills required to become an offshore HDA
Learning outcomes	Role and responsibilities of an HDA, offshore helideck legislation, equipment and systems relevant to helideck operations, helicopter refuelling and associated helideck hazards, providing assistance for cargo and passenger handling on helidecks
Assessment	Assessment is carried out by direct observation and oral and/or written questions
Duration	18 hours
Validity of certificate	No expiry
Refresher training	NA

11.3.15 HDA Helideck Emergency Response Team Member (HERTM) Training

Target group	**Helideck assistants requiring HERTM training to perform as emergency response team member**
Pre-requisite training	None

Medical fitness	Offshore medical certificate or medical screening by the training provider
Training objectives	To gain knowledge and skills required for helicopter emergency response as helideck team member
Learning outcomes	Role and responsibilities of HDA HERTM, understand the helideck emergency response plan, identify helideck emergencies, understand helideck emergency systems, know the difference in emergency response between manned and unattended installations, communications, BA operations, fire fighting on helideck using fixed and mobile equipment, deal with non-fire helideck emergencies, SAR operations, handling casualties
Assessment	Demonstration in practical exercises combined with oral and/or written questions
Duration	26 hours
Validity of certificate	2 years
Refresher training	HDA Further training

11.3.16 HDA Further Training

Target group	**Personnel tasked to assist on helideck during helicopter operations who need to revalidate their HDA certificate**
Pre-requisite training	HDA Helideck Emergency Response Team Member Training and HDA Initial Training
Medical fitness	Offshore medical certificate or medical screening by the training provider
Training objectives	To refresh knowledge and skills required for helicopter emergency response as helideck team member
Learning outcomes	Remaining calm when responding to emergencies, asking for assistance, communication with HLO and other team members, execution of emergency response tasks, correct use of SCBA, helideck SAR techniques, use of fire fighting systems and casualty handling
Assessment	Demonstration of skills in practical exercises
Duration	12 hours 30 minutes
Validity of certificate	2 years
Refresher training	HDA Further training

11.3.17 HLO Training

OPITO HLO training is divided into six phases as given below:

1. HLO initial training
2. HERTM training
3. HERTL training
4. HLO workplace experience and competence assessment
5. HLO further training
6. HLO workplace drills and competence reassessment – workplace assessment and reassessments are conducted on-board the offshore installation or the vessel as it requires site specific training and assessment. Details of the units assessed offshore are

given in the OPITO HLO standard. This training consists of the following units:

Unit 1:   Maintaining a State of Readiness
Unit 2:   Supervise Helicopter Landing and Departure
Unit 3:   Supervise Cargo Handling
Unit 4:   Supervise Refuelling the Helicopter
Unit 5:   Control the Response to Emergencies.

Details of the HLO initial training are given below whilst other shore based courses are described in the next subsections.

11.3.17.1 HLO Initial Training

Target group	**Personnel assigned or likely to be assigned the role of Helicopter Landing Officer**
Pre-requisite training	None
Medical fitness	Offshore medical certificate or medical screening by the training provider
Training objectives	To provide theoretical and practical knowledge to perform the role of HLO
Learning outcomes	Legislation related to helideck operations; hazards, equipment, systems and characteristics of helidecks, responsibilities of HLO, actions required by HLO prior to helicopter arrival, during and after landing, weather reports required by helicopter pilot; identification, packaging and labelling of dangerous goods, HLO requirements for normally unattended installations (NUI), radio communications, inspection of helideck after takeoff, checking of passenger and freight manifests, supervision of HDAs
Assessment	Demonstration of skills in practical exercises combined with oral and/or written questions
Duration	18 hours
Validity of certificate	2 years
Refresher training	Helideck Emergency Response Team Leader (HERTL)

11.3.17.2 HLO Helideck Emergency Response Team Leader (HERTL) Training

Target group	**Personnel currently appointed as HLOs who require HERTL training as part of their overall competence requirement**
Pre-requisite training	HLO Initial Training and HERTM
Medical fitness	Offshore medical certificate or medical screening by the training provider
Training objectives	To provide detailed knowledge of helideck emergency response to HLOs so that they can take the role of HERTL
Learning outcomes	Roles and responsibilities of HLO and HLO HERTL, helideck emergency response plan, types of helideck emergencies, fire fighting safety, helideck emergency systems and controls, difference in emergency response between manned and Normally Unmanned Installations (NUI), effective radio communications, use of BA, control of HERT, managing missing persons and recovering casualties, dealing with stress in HERT

Assessment	Demonstration of skills in practical exercises combined with oral and/or written questions
Duration	26 hours (20% theory, 80% practical)
Validity of certificate	2 years
Refresher training	HLO Further training

11.3.17.3 HLO Further Training

Target group	**Personnel currently appointed as HLOs whose certification requires to be revalidated**
Pre-requisite training	HLO Further Training or HERTL Certificate
Medical fitness	Offshore medical certificate or medical screening by the training provider
Training objectives	To refresh knowledge of helideck emergency response to HLOs for areas that cannot be covered offshore
Learning outcomes	Fire fighting safety, helideck emergency systems and controls, effective radio communications, monitoring use of BA, controling HERT, managing missing persons and recovering casualties, dealing with stress in HERT
Assessment	Demonstration of skills in practical exercises combined with oral and/or written questions
Duration	12.5 hours (10% theory, 90% practical)
Validity of certificate	2 years
Refresher training	HLO Further training

11.3.18 Major Emergency Management Initial Response (MEMIR) Training

Target group	**Persons in charge or members of or providing support to the emergency management team**
Pre-requisite training	None
Medical fitness	Offshore medical certificate or medical screening by the training provider
Training objectives	To provide knowledge and skills to prepare for, respond to and maintain control of an emergency situation, manage communication, put plans into actions as dictated by the situation, monitor stress during emergencies
Learning outcomes	Review, manage and assess available information, take effective actions, implement emergency plans, communicate effectively including appropriate agencies with support from emergency response team, manage resources, adapt plans as required, deal with stress and delegate authority
Assessment	Demonstration of skills in practical exercises combined with oral and/or written questions
Duration	26 hours (35% theory, 65% practical)
Validity of certificate	NA
Refresher training	NA

11.3.19 Offshore Emergency Response Team Leader (OERTL) Initial Training

Target group	**Personnel tasked or likely to be tasked the role of an OERTL**
Pre-requisite training	Offshore Emergency Response Team Member (OERTM) Certificate
Medical fitness	Offshore medical certificate or medical screening by the training provider
Training objectives	To provide knowledge and skills to become a leader of the Offshore Emergency Response Team
Learning outcomes	Knowledge required for leadership including relevant legislation. Monitor self and other team member's stress, monitor BA use and other topics similar to MEMIR
Assessment	Demonstration of skills in practical exercises combined with oral and/or written questions
Duration	24 hours, 20% theory 80% practical
Validity of certificate	2 years
Refresher training	OERTL Further Training

11.3.20 Offshore Emergency Response Team Leader (OERTL) Further Training

Target group	**Personnel holding an OERTL certificate**
Pre-requisite training	Offshore Emergency Response Team Leader certificate or OERTL Further Training certificate
Medical fitness	Offshore medical certificate or medical screening by the training provider
Training objectives	To update and refresh the skills and knowledge that cannot be refreshed when working offshore
Learning outcomes	Effective communications, managing emergency response team, briefing and debriefing teams, monitoring stress in team members
Assessment	Demonstration of skills in practical exercises combined with oral and/or written questions
Duration	10 hours, 10% theory 90% practical
Validity of certificate	2 years
Refresher training	OERTL Further Training

In addition to the above shore based training for offshore emergency response i.e. OERTL initial/further training, company and site specific training will be carried out at the workplace for which OPITO standard specifies the requirements.

11.3.21 Offshore Emergency Response Team Member (OERTM) Initial Training

Target group	**Personnel assigned or likely to be assigned the role of an emergency response team members**
Pre-requisite training	BOSIET/T-BOSIET, FOET/T-FOET
Medical fitness	Offshore medical certificate or medical screening by the training provider
Training objectives	To provide knowledge and skills to personnel so that they can perform the role of an OERTM
Learning outcomes	Role of OERTM, purpose of emergency response, fixed fire systems and fire fighting, PPE requirements, under directions from OERTL take actions to assess the emergency, communicate effectively with OERTL and other team members, BA operations, SAR for missing persons and recover casualties
Assessment	Demonstration of skills in practical exercises combined with written questions
Duration	32 hours, 25% theory 75% practical
Validity of certificate	2 years
Refresher training	OERTM Further Training

11.3.22 Offshore Emergency Response Team Member (OERTM) Further Training

Target group	**Personnel holding an OERTM certificate**
Pre-requisite training	Offshore Emergency Response Team Member (OERTM) initial training certificate or OERTM Further Training certificate
Medical fitness	Offshore medical certificate or medical screening by the training provider
Training objectives	To update and refresh the skills and knowledge that cannot be refreshed when working offshore
Learning outcomes	Enter and work in live fires, locate and recover casualties, use of appropriate portable fire fighting equipment, BA operations, responding to confined space and working at height incidents, effective communications
Assessment	Demonstration of skills in practical exercises combined with oral and/or written questions
Duration	10 hours, 10% theory 90% practical
Validity of certificate	2 years
Refresher training	OERTM Further Training

In addition to the above shore based training for offshore emergency response i.e. OERTM initial/further training, company and site specific training will be carried out at the workplace for which OPITO standard specifies the requirement.

11.3.23 Offshore Muster Checker

Target group	**Personnel designated the role of an offshore muster checker. Training conducted by the duty holder on-board installation**
Pre-requisite training	BOSIET/FOET
Medical fitness	Offshore medical certificate or medical screening by the training provider
Training objectives	To enable personnel to undertake the role of a muster checker
Learning outcomes	This training consists of two units: 1. Maintain readiness for muster, evacuation and respond to emergency 2. Conduct physical headcount, maintain control and respond to instructions
Assessment	The assessment will be carried out by a competent assessor according to the OPITO guidance.
Duration	NA
Validity of certificate	NA
Refresher training	Participation in regular drills offshore

11.3.24 Offshore Muster Coordinator

Target group	**Personnel designated the role of an offshore muster coordinator. Training conducted by the duty holder on-board installation**
Pre-requisite training	BOSIET/FOET
Medical fitness	Offshore medical certificate or medical screening by the training provider
Training objectives	To enable personnel to undertake the role of a muster checker coordinator
Learning outcomes	This training consists of three units: 1. Maintain readiness for muster and evacuation 2. Control muster and coordinate evacuation 3. Maintain effective communications
Assessment	The assessment will be carried out by a competent assessor according to the OPITO guidance
Duration	NA
Validity of certificate	NA
Refresher training	Participation in regular drills offshore

11.3.25 Offshore Radio Operator during Emergencies Initial Training

Target group	**Personnel assigned or likely to be assigned the role of offshore radio operator during emergencies**
Pre-requisite training	CAA issued offshore radio operator's certificate of competency or equivalent
Medical fitness	Medical screening by the training provider
Training objectives	To provide knowledge of communications during emergency response as related to the offshore radio operator to enable them to carry out effective communications
Learning outcomes	Role of radio operator, emergency response arrangements, applicable legislation, maintaining state of readiness
Assessment	Demonstration of skills in practical exercises combined with 30 minutes written test
Duration	8 hours
Validity of certificate	NA
Refresher training	NA

11.3.26 OIM Controlling Emergencies

Target group	**Personnel likely to be in command of an offshore installation**
Pre-requisite training	Element 1.1 of the OPITO standard must be completed prior to attending this training. Delegates are recommended to have completed Major Emergency Management Initial Response course.
Medical fitness	Medical screening by the training provider
Training objectives	To assess delegates' competency for controlling offshore emergencies
Learning outcomes	Assessment of delegates will be based on at least three emergency scenarios where OIM will be required to take appropriate actions to control them. The certificates can be endorsed with any of the following:

- Production Operations (Fixed)
- Production Operations (Floating)
- Drilling Operations (Fixed)
- Drilling Operations (Floating)
- Mobile/Floating Installations
- Normally Unattended Installations (NUI)

The scenarios to choose from are:

- Accommodation fire
- Collision or wave damage causing structural collapse
- Explosion and fire
- Helicopter incident
- Loss of stability (mobile installations)
- Pipeline incident
- Well control incident

Assessment	Observation of performance by two assessors during simulation based exercises
Duration	Dependent upon requirements
Validity of certificate	3 years (Oil & Gas UK guideline)
Refresher training	OIM Controlling emergencies

11.3.27 Minimum Industry Safety Training (MIST)

Target group	**Personnel new to the offshore oil and gas industry**
Pre-requisite training	None
Medical fitness	Offshore medical certificate or medical screening by the training provider
Training objectives	To introduce attendees to the key safety elements required for all offshore workers
Main topics covered	Major Accident Hazards, Workplace Hazards and Personal Safety, Risk Management, Control of Work, Helicopter Safety
Assessment	Demonstration of skills in practical exercises combined with a written test at the end of each module
Duration	13 hours 30 minutes
Validity of certificate	4 years
Refresher training	Online assessment

11.3.28 International MIST

Target group	**Personnel new to the offshore oil and gas industry (other than in the UK where MIST course is required)**
Pre-requisite training	None
Medical fitness	Offshore medical certificate or medical screening by the training provider
Training objectives	To introduce attendees to the key safety elements required for all offshore workers
Main topics covered	Introduction to the hazardous offshore environment Working safely including safety observations systems, risk assessment, tasks that require permit to work, asset integrity, manual handling, control of substances hazardous to health (COSHH), working at height and mechanical lifting
Assessment	Demonstration of skills in practical exercises combined with a written test at the end of each module
Duration	13 hours 30 minutes
Validity of certificate	4 years
Refresher training	Online assessment

11.4 Ongoing On-board Development and Training Programme for ERRV Masters and Crews (OODTP)

This training programme is for on-board development and training for ERRV masters and crews. The assessors visit the vessels at intervals not exceeding 12 months where assessment for each crew member must be carried out at least once every three years. Assessors may check the appraisal reports and comments from the vessel and the assessors. The assessment may also include evidence provided by the vessel operator or duty holders. Certificates for candidates who are unable to demonstrate competence in the assessment or who fail to attend the assessment will revert to the next low level certificate until they retake the training necessary to reach the previous level. For example, a daughter craft coxswain who allows their certificate to lapse will be required to undertake FRC Boatman, FRC Coxswain and then daughter craft coxswain. The system works through OPITO's central register (Vantage) where assessors send reports to update the system.

11.5 Industry Training and Competency Standards

In addition to the emergency and critical response standards described in the previous section, OPITO has also published Industry Training and Competency Standards for areas that may be similar to the shore based industries but require additional specific consideration for the offshore work environment. Full details of these standards can be found on OPITO's website at www.opito.com/standards-library but a list of these standards is given here for reference.

1. Application of Insulation Systems Competence Assessment (Level 2)
2. Application of Insulation Systems Training (Level 1)
3. Authorised Gas Tester Training Level 1 – Further testing for confined space
4. Authorised Gas Tester Training Level 1 – Initial testing for confined space
5. Authorised Gas Tester Training Level 2 – Further testing for Hot Work
6. Authorised Gas Tester Training Level 2 – Initial testing for Hot Work
7. Authorised Gas Tester Training Level 3 – Further Gas Monitoring for Hotwork Sites

8. Authorised Gas Tester Training Level 3 – Initial Gas Monitoring for Hotwork Sites
9. Banksman and Slinger Competence Stage 3
10. Banksman and Slinger Competence Stage 4
11. Banksman and Slinger Training – Stages 1 and 2
12. Blaster – Sprayer Competence Standard
13. Blaster/Sprayer Training (Level 1)
14. Competence Assessor Training Standard
15. Control of Work Refresher Training for Performing Authorities
16. Control of Work Training for Performing Authorities
17. Elected Safety Representatives Development Training
18. Internal Verifier Training Standard
19. LOLER Competent Person
20. Offshore Crane Operator – Stage 1 Introductory Training
21. Offshore Crane Operator – Stage 2 Competence Assessment Standard
22. Offshore Crane Operator – Stage 2 Competence Reassessment Standard
23. Offshore Crane Operator – Stage 2 Training
24. Offshore Crane Operator – Stage 3 Competence Assessment Standard
25. Offshore Crane Operator – Stage 3 Competence Reassessment Standard
26. Offshore Safety Representatives Further Training
27. Offshore Safety Representatives Initial Training
28. Oil and Gas Industry Fireproofing Training (Level 1)
29. Oil and Gas Industry Fireproofing Competence
30. Preparation of Dangerous Goods for Transport by Sea – Initial and Refresher Training
31. Rigger Competence Stage 3
32. Rigger Competence Stage 4
33. Rigger Training (Stages 1 and 2)
34. Safe Driving at Work Standard

11.6 OPITO Certification – Validity and Dispensations

Due to the nature of offshore work and means of 'weather' dependent transportation, at times it may not be possible to return in time to a shore based training provider to undertake refresher courses. In order to provide completeness and facilitate the workers, Oil & Gas UK have agreed a mechanism to provide 'dispensations' for 'emergency response training' courses where workers can't undertake refresher courses before the expiry of their certificates. UK Oil & Gas's guidance and the information required to be completed is given in the 'Request for Dispensation form' that follows. The latest version of this form can be downloaded from www.oilandgasuk. co.uk/.

Personnel wishing to obtain the benefit of dispensation must ensure that the completed form is signed by the duty holder without which it will not be considered valid and the training provider will be unable to authorise dispensation.

Other important factors to keep in mind when considering dispensations and expiry dates are given below:

- In most of the heliports, workers will not be allowed to travel offshore without valid certificates. This can be checked through individuals producing the original certificate at the heliport or by access to the Vantage database where training details for all workers are updated upon completion of training. However, in some cases where workers need to travel on the day immediately after completion of training, the information may not have been uploaded to Vantage. In such cases, workers are advised to take their original certificate with them to avoid any last minute complications.
- FOET course can be undertaken up to three months PRIOR to or AFTER the expiry date of BOSIET or FOET. However, in order to maintain consistency, the expiry date of the original certificate will be maintained for the new certificate.
  - *Back Dated Certificates*: If an individual's FOET course certificate expired on 1 March 2015, they obtained a dispensation and managed to attend the course on 1 April 2015, then the new FOET certificate will still have an expiry date of 28 February 2019.
  - *Forward Dated Certificates*: If an individual whose FOET certificate expires on 1 May 2015 attends the course on 1 March 2015, then the new certificate's expiry date will be forward dated to 30 April 2019.
- For Coxswain or Emergency Response Fire Training, there is a dispensation period of 3 months for attending prior to the expiry of original certificate.

11.7 Mutual Recognition of Training

The majority of the workforce in the EU and elsewhere in the world are likely to move between different regions. Whilst OPITO has been working on globalisation of its standards, there are many countries around the world, especially in the North Sea sector, which maintain their own standards for training.

Whilst there may be difference in some training, the standards for Emergency Preparedness training are similar in Norway, Denmark, Netherlands and the United Kingdom. In order to avoid duplication of training, representatives from Oil & Gas UK, Norsk

Appendix 1

Request for Dispensation
(Extension to an expired emergency response training certificate)

Notes:

All reasonable efforts must be made to ensure that emergency response related refresher training is completed before the individual's current certificate expires.

In exceptional circumstances, where it is not possible for an individual to complete a refresher training course before their current refresher certificate expires, it may be possible for the Duty Holder of the Installation on which the individual works, or is due to work, to extend the currency of the individual's current training certificate by up to 3 months. Such extensions should only be granted to cover unforeseen circumstances such as illness or abnormal work demands.

- o Extensions will only be applied to personnel who work offshore regularly and not to visitors or personnel who work offshore occasionally.
- o Extensions will only be applied to personnel who have completed basic training and at least one related refresher course in the respective emergency response subject matter
- o The 'Responsible Person' named below must be nominated by the Duty Holder and will normally be an OIM or senior member of the onshore operational staff. The Responsible Person should review the individual's training record, experience and frequency of participation in relevant drills and exercises before determining that attendance at a refresher course is adequate. Otherwise, the individual should complete the relevant basic training course again.
- o Where an extension is granted, the effective start date of the new refresher training certificate will be the expiry date of the individual's corresponding, current certificate.

Name of person requiring dispensation:		
Date of Birth:	Vantage number:	
Training Course name:		
Current Certificate Expiry Date:		
Organisation requesting dispensation:		
Responsible Person:		
Job Title:		
Contact telephone number:		
Contact email address:		
Reason for granting dispensation:		
Responsible Person Signature:	Date:	

For your information: Please also note that valid 'Emergency Response' Training Certificates can be 'refreshed' up to 3 months prior to the existing certificate expiry date (without loss of validity).

OGUK Dispensation Form V1 November 2009

Figure 11.1 Oil & Gas UK's form for requesting dispensation

Table 11.1 – Mutual recognition of Basic and Refresher Safety and Emergency Preparedness Training (Revised 01.02.2013)				
Training From ↓	Working In →	Norway (Norsk Olje & Gas)	Denmark (NSOC-D)	Netherlands (NOGEPA)	United Kingdom (Oil & Gas UK)
Norway (Norsk Olje & Gas)			Accepted	Accepted	Accepted
Denmark (NSOC-D)		Accepted with additional escape chute training		Accepted	Accepted with additional MIST training
Netherlands (NOGEPA)		Accepted with additional escape chute training	Accepted		Accepted
United Kingdom (Oil & Gas UK)		Accepted with additional escape chute training	Accepted	Accepted	

Abbreviations:
MIST: Minimum Industry Safety Training
NOGEPA: Nederlandse Olie en Gas Exploratie en Productie Associatie

Norsk Olje & Gas: Norwegian Oil and Gas Association
NSOC-D: North Sea Operators Committee – Denmark

Olje & Gas in Norway, NSOC-D in Denmark and NOGEPA in the Netherlands agreed to recognise each other's training to pass benefits to the workforce working across borders in the North West Europe. A matrix of training recognition is given in Table 11.1.

11.8 Medical Standards for Work in Offshore Oil and Gas Industry

Workers in the offshore oil and gas industry are required to be physically and mentally fit to travel offshore and undertake their jobs. In the UK, Oil & Gas UK publishes guidelines for these requirements.

Until 2007, all workers were required, depending upon their age, to be assessed as below:

- Age under 40 – every 3 years
- Age 40–49 – every 2 years
- Age 50 or above – every year.

However, the above requirements changed to a frequency of two years for all workers regardless of their age. The assessing physician/doctor may, however, change this frequency depending upon an individual's medical condition.

Assessing physicians, doctors or medical practitioners have to be registered with Oil & Gas UK to carry out required assessments and issue offshore medical certificates. Their list and contact details can be found on Oil & Gas UK's website at www.oilandgasuk. co.uk/knowledgecentre/doctors.cfm.

Note

1 OPITO (2014) Missions and Objectives. [Online] www.opito.com/
content/mission-objectives [Accessed 12.12.14]

Bibliography

Oil & Gas UK [Online] www.oilandgasuk.co.uk
OPITO Standards Library [Online] www.opito.com/Standards-
library

Index

Page numbers in *italics* denote an illustration, **bold** indicates a table